RAND NATIONAL DEFENSE RESEARCH INSTITUTE

Policies for Managing Reductions in Military End Strength

Using Incentive Pays to Draw Down the Force

Michael G. Mattock, James Hosek, Beth J. Asch

Prepared for the Office of the Secretary of Defense

For more information on this publication, visit www.rand.org/t/RR545

Library of Congress Cataloging-in-Publication Data
ISBN: 978-0-8330-9231-1

Published by the RAND Corporation, Santa Monica, Calif.

Preface

Policy changes periodically call for military force downsizing. A major decrease in force size occurred at the end of the Cold War, the Navy downsized in the previous decade, and the Army and Marine Corps are currently downsizing. Force size can be cut by decreasing the number of new recruits, allowing early departures, tightening physical and promotion standards, and denying continuation or reenlistment, for instance. But if cuts are not made carefully, the resulting force can have imbalances, and they can persist for years and create future management challenges. Also, when cuts are deep, personnel who have performed well can face involuntary separation, and this could decrease morale and weaken a service's reputation for loyalty to its members. This report develops a drawdown tool, voluntary separation pay, designed to avoid these adverse effects. Voluntary separation pay can achieve a drawdown of a given size in a given period of time and do so at least cost and in a way producing the desired ex post experience mix. In contrast, a policy of involuntary separations can impose direct and indirect costs on service members. Involuntarily separated members could be undercompensated relative to what would be required for voluntary separation, and retained members might perceive a heightened risk of involuntary separation at some future date. This report should interest decisionmakers, analysts, and researchers concerned about military compensation and the retention or separation of military personnel.

This research was sponsored by Cost Assessment and Program Evaluation in the Office of the Secretary of Defense and conducted within the Forces and Resources Policy Center of the RAND National Defense Research Institute, a federally funded research and development center sponsored by the Office of the Secretary of Defense, the Joint Staff, the Unified Combatant Commands, the Navy, the Marine Corps, the defense agencies, and the defense Intelligence Community.

For more information on the RAND Forces and Resources Policy Center, see www.rand.org/nsrd/ndri/centers/frp or contact the director (contact information is provided on the web page).

Contents

Figures and Tables

Figures

Tables

Summary

Well-designed voluntary separation pay (VSP) can be used to decrease a military personnel force voluntarily, promptly, and without lingering adverse effects on personnel force structure, such as personnel imbalances by year of service (YOS). The use of VSP requires an initial increase in personnel expenditures but produces net decreases in personnel costs in subsequent years.

Approach

We extended RAND's dynamic retention model (DRM) to design separation incentives, called VSPs, to decrease Army active component end strength without creating imbalances in experience. The DRM shows how retention changes when pay changes, and it can be used to simulate the effect of separation pays of different amounts and targeted to different years of service, as in a conventional what-if analysis. Going beyond this, we used the DRM to solve directly for the amount of separation pay required to induce a given percentage of service members at a particular YOS to separate voluntarily. We combined the results across YOS cohorts to find a schedule of VSPs that scale down the force to a target level. By keeping the decreases by YOS in scale, it was possible to achieve a smaller force with the same relative experience mix as the original force. The resulting force therefore did not have an experience gap or "bathtub," and it maintained opportunities for promotion and the progression of personnel through the ranks as they accumulated experience. We can adapt the method to reach other experience mixes.

We looked at a variety of drawdown scenarios and, through policy simulation, calculated the VSPs required to achieve the drawdowns, the cost to the Army of implementing the VSPs, the decrease in personnel costs resulting from the departure of personnel, and the experience-profile effects of VSPs over time under each scenario. The scenarios included offering VSPs for one, two, or three years; determining the VSPs needed for an across-the-board cut in force size versus cutting portions of the force more deeply; and determining the effect that announcing VSPs ahead of time had on VSP. The research focused on developing VSP as a drawdown tool, and the report discusses VSP relative to other approaches.

Key Findings

- VSP can draw down the force rapidly without creating an imbalance in the experience mix of personnel (a bathtub).

- A higher VSP is need to make a deeper cut in force size, but net personnel cost decreases even so.
- A lower VSP offered for multiple years can achieve the same end-strength goal as a VSP offered for one year, but net decreases in personnel costs are lower because more personnel remain on hand while the VSP is offered.
- If members anticipate being offered VSP, a higher VSP is required to reach a given target.
- A 10-percent cut in the force would produce a net decrease in the present value of active component Army personnel costs of $6.4 billion to $7.4 billion in 2013 dollars over the first ten years.
- The cost of implementing VSPs is $1.7 billion to $3 billion for a 10-percent cut in the force, depending on whether the VSP is implemented for a window of one, two, or three years and whether the cut is across the board or concentrated on a portion of the force.
- VSPs can be targeted on particular groups, e.g., poorer performers or personnel in over-strength occupations, and thereby induce separations from those groups.

VSPs can draw down the force evenly to a level close to the desired steady-state size and shape (experience mix) within a relatively short period of time, e.g., a year, with no follow-on effects requiring management in later years. Figure S.1 shows the effect of VSPs offered to officers in YOS 7 to 18 combined with an involuntary separation policy for YOS 1 through 6. The VSPs are designed to draw the force down by 10 percent within three years. By the end of three years, the officer population is at or near the steady state for the desired end strength in YOS 1 through 21, and the excess officer population beyond YOS 20 is only 0.9 percent of the original strength. The excess population seems manageable by tighter promotion standards.

Although larger VSPs are required for larger decreases in end strength, a net decrease in personnel costs can still be realized. Even cuts as deep as 40 percent in selected military occupational specialties can produce a net decrease in personnel costs within a few years. This is well within the six-year window of the Future Years Defense Program.

The net decrease in personnel costs is lower if VSP is offered for multiple years. The smaller net decrease in personnel costs occurs because, although the VSP is smaller, personnel leave more gradually, so personnel costs do not decrease as rapidly. However, a more gradual decrease increases the potential for regenerating the force and spreads the cost of implementing the VSP over several years.

The net decrease in personnel costs is also less if members can anticipate that a VSP will be offered. Higher VSPs will be needed because members are more likely to choose to stay in anticipation of the future availability of the VSP. The increase in the required VSP is greater in earlier YOS; members in later YOS are unlikely to leave in any case, and the future availability of a VSP has little effect on their behavior.

The net decrease in personnel costs for Army officers and enlisted personnel is $6.4 billion to $7.4 billion in 2013 dollars in the first ten years of scenarios in Table S.1. Each of the scenarios decreases end strength by 10 percent. The first three vary the length of time the VSP is offered from one to two to three years. The other scenarios offer VSP for one year and consider cuts of 10 to 40 percent. In the "draw down all to 90%" scenario, the baseline force is cut by 10 percent, and the other two scenarios are cuts of 20 percent to half the force and 40 percent to one-quarter of the force.

Tables S.2 and S.3 show the initial cost of implementing VSPs in the first year and the decrease in personnel costs ("compensation costs avoided") in the following years. Depending

Acknowledgments

We are grateful to M. Webster Ewell, Office of the Secretary of Defense Cost Assessment and Program Evaluation, for suggesting this topic and for his guidance and support throughout the research. We thank our RAND colleagues John D. Winkler, Igor Mikolic-Torreira, and Michael L. Hansen for their thoughtful comments during the course of the study on an earlier draft of this report.

Abbreviations

AC	active component
DRM	dynamic retention model
ISP	involuntary separation pay
NCP	normal cost percentage
OMB	Office of Management and Budget
PV	present value
RC	reserve component
RMC	regular military compensation
SSB	special separation bonus
TERA	Temporary Early Retirement Authority
VSI	voluntary separation incentive
VSP	voluntary separation pay
YOS	year of service

CHAPTER ONE

Introduction

The Office of the Secretary of Defense Cost Assessment and Program Evaluation asked the RAND National Defense Research Institute to devise a monetary incentive for voluntary separations that could be used for military drawdowns, with a particular focus on the Army. The incentive would need to be effective in reaching a given decrease in force size within a given period of time, achieve the desired post-drawdown experience mix of personnel, not cause an outflow of high performers, and be cost-efficient. The voluntary separation pay (VSP) developed in this research meets these objectives.

To design VSP, we made use of the dynamic retention model (DRM) capability developed at RAND in the past decade. The DRM can predict the retention response to a monetary incentive at an individual level and aggregate across individuals to show how the experience mix of retained personnel evolves during and after the drawdown. The DRM has been extensively used to analyze possible changes to special and incentive pays, reserve compensation, and military retirement. Specific challenges in the present research were to find the least-cost VSP for a given drawdown, to consider how the amount of VSP changed when it was to be offered for a single year or multiple years and when it targeted a deeper cut for part of the force versus a shallower cut of the entire force, and to determine whether it was more or less cost-effective when it was announced ahead of time and personnel could anticipate its availability. In addition, the analysis needed to show the cost of implementing the VSP and the personnel costs avoided in subsequent years as personnel left the force.

This document summarizes the issues, methods, simulations, and findings. Chapter Two is a precursor to the analysis and discusses policies that can be used to draw down the force. Chapter Three describes the analytical approach and introduces the DRM, and there is a technical discussion of it in the appendix. Chapter Four describes the results of the analyses, presenting VSPs for enlisted members and officers, by year of service (YOS), for drawdowns of various depths and various approaches to implementation (different window lengths, taking a given cut in force size from the overall force or a portion of it, and announcing the availability of VSP ahead of time, or not). Chapter Four also shows the cost of implementing VSPs for each scenario and the net savings in personnel costs in the following years. Chapter Five summarizes the key findings of the analysis and simulations and discusses voluntary separation with VSP versus the alternative used today, involuntary separation with involuntary separation pay (ISP).

CHAPTER TWO

Policies to Achieve a Drawdown

Policies to decrease force size include personnel actions and separation incentives. The personnel actions are decreased accessions—the intake of new personnel—of new service members, early separation, denial of continuation or reenlistment, tighter physical standards, and tighter promotion standards. The separation incentives are ISP, voluntary separation incentive (VSI), special separation bonus (SSB), and VSP. VSI and SSB were discontinued in 2001, but we include them for completeness. The personnel policies can be used at the same time. Also, ISP and VSP can be used at the same time. ISP would be paid to members not meeting the physical or promotion standards and therefore not eligible to continue, and VSP would be paid to members eligible to continue yet meeting VSP eligibility criteria. The criteria could aim at lower-performing personnel.

Decreased Accessions

By decreasing accessions, force size can be decreased over a period of years until a new, lower force size is achieved. This saves costs by decreasing recruiting resources, such as recruiters, recruiting stations, advertising, enlistment bonuses, and educational benefit supplements. But there are limits to the usefulness of this approach in the military. If accessions decrease below the steady-state level for the new, lower force size, there will be a bathtub as these small cohorts move through their military careers. If accessions decrease by less than the steady-state level for the new force size, steps must be taken later to trim these cohorts to size. Further, if only a decrease in accession is used, the experience mix of personnel will be off kilter, with too few junior personnel relative to senior personnel.[1] Finally, to the extent that accessions are decreased, the positions of incumbent personnel are protected. This is advantageous to these personnel but not necessarily advantageous to the service. It must continue pay the personnel costs of incumbent personnel, which is higher than that for new entrants.

[1] Suppose the active component (AC) enlisted Army had 490,000 soldiers with an average length of stay of seven years and wanted a 10-percent decrease. If the force were in a steady state, it would require (490,000 ÷ 7 =) 70,000 accessions. A 10-percent decrease would be 49,000 soldiers. In the new steady state, accessions would be 10 percent lower, or 7,000 per year lower, and a cut of 49,000 soldiers would take seven years if done by accessions alone. The drawn-down force would have a junior portion sized for the new, smaller force and a senior portion based on the old, larger force. This would probably be an undesirable experience mix.

Early Outs with Voluntary Departure

Enlisted members and officers can be allowed to leave before completing their terms of service or service obligations. Those choosing to leave early would be those who expected to leave at the end of their obligations. By allowing early departure, the Army can free up spaces that need not be refilled, which is a step in downsizing. The personnel opting to depart do so voluntarily and do not need to be paid to leave. Early outs could be offered in a limited way, e.g., to enlisted members within a year of the end of their first-, second-, or third-term obligations and to officers within a year or 18 months of their active-duty service obligations.

Denial of Continuation or Reenlistment

Under a reduction in force, a member is denied continuation or retention. An analogue is when a company closes a plant and workers are left without jobs (displaced), which, of course, is different from being fired for cause. Reductions in force can be targeted to functional areas, occupations, YOS groups, or lower-performing personnel. A reduction in force need not designate specific individuals for separation but can operate impersonally by setting a departure goal and offering separations on a first-come, first-served basis. This can be expected to attract personnel who value their military careers less than their external opportunities. This could include high-performing personnel—hence the importance of setting criteria that focus exit opportunities on lower performers.

Tighter Physical Standards

Although standards can be tightened somewhat without causing concern, the usefulness of tighter standards depends more generally on the validity of evidence linking the standard to performance on duty. The extent to which a standard would need to be changed depends on the size of the drawdown and the number of personnel near the standard. If many personnel are near the standard, a small change might disqualify enough personnel to reach the target. A small or large change could be viewed as an arbitrary approach to separating personnel if members believe that the standard has low validity, i.e., believe that personnel near the standard perform as well as personnel above the standard. Low validity could be the case in occupations requiring knowledge and experience but having few tasks for which physical strength and endurance are critical. If strength and endurance are critical but many members are well above the standard, a large change in the standard would be needed, again perhaps calling into question its validity. Tighter standards also could induce behavioral changes. Personnel who want to stay in the military but anticipate being ineligible under the tighter standard could improve their physical condition. Those who want to leave could let their condition deteriorate.

Tighter Promotion Standards

For enlisted members, tighter promotion standards take the form of decreasing the allowable time in grade and increasing competitiveness standards. When less time is allowed and competitiveness standards are higher, more service members will be ineligible to continue in service. Officers twice passed over for promotion are ineligible to continue beyond completion of the existing service obligation.

A lower probability of promotion is related to, but different from, tighter promotion standards. When the pool of service members promotable to the next grade is larger than the number of available positions, the probability of ever being promoted to the next grade is lower. This could happen if the service is downsizing, and downsizing could also prompt tighter time-in-grade standards for enlisted members and lower selection probabilities for officers. A service's decision to decrease the number of higher-grade positions is likely to be common knowledge within the service, and members can foresee a slower speed of advancement or lower chance of promotion. This decreases the value of staying in the military and could induce personnel to leave, especially those who feel that their chances of advancement are low. If the promotion system is working to select higher performers for advancement, a decrease in promotion speed and opportunity would be proselective on quality in that lower performers would be more adversely affected and more likely to leave.

Involuntary Separation Pay

U.S. Department of Defense Instruction 1332.29 (Under Secretary of Defense for Personnel and Readiness, 1991 [2011]) states that ISP is "authorized to members of the Regular and Reserve components involuntarily separated" from active duty who meet all of the following conditions: (1) the service member has completed at least six but fewer than 20 years of active service, (b) the separation is "honorable," and (c) the service member is being involuntarily separated "through either the denial of reenlistment or the denial of continuation on active duty" because of a specific condition, including "promotion or high year of tenure policies" or "reduction in force."[2] So, if a member exceeds time in grade for promotion, the service no longer has a space for his or her specialty and level, or the member is denied reenlistment or continuation, the member is involuntarily separated and paid ISP. The Department of Defense policy does not identify criteria under which a member can be denied reenlistment or continuation. ISP is a lump-sum payment equal to 10 percent of the annualized value of the member's current monthly basic pay times the number of YOS, including fractions of years—that is, 0.1 × 12 × current monthly basic pay × YOS. A member with ten years of service will receive one year's worth of basic pay, for instance.

Voluntary Separation Incentive and Special Separation Bonus

These pays were offered in the 1990s and discontinued in 2001. Temporary Early Retirement Authority (TERA), instituted in the National Defense Authorization Act for Fiscal Year 2012 (Pub. L. 112-81), provided another type of retirement pay that was discontinued in 2002. A 2012–2018 TERA program was created and is expected to end in December 2018.

---

[2] The instruction also lists other conditions that relate primarily to reserve forces. The two conditions given in the text are the most relevant for our discussion.

VSI was an annuity calculated as 2.5 percent of the member's basic pay times YOS and paid for a period twice as long as the member's active YOS. SSB was a lump sum calculated as 15 percent of the member's annualized monthly basic pay times YOS. The member could choose either VSI or SSB; those with a real personal discount rate of about 20 percent would be indifferent to SSB's lump sum versus VSI's annuity stream. Using the VSI/SSB choice, Warner and Pleeter, 2001, estimated personal discount rates ranging from 0 to 30 percent and where the rate varied with demographic characteristics, education, test scores, and size of payment. According to DRM estimates, the personal discount rate is 10 to 12 percent for enlisted members and 6 percent for officers. At these rates, the VSI annuity would be preferred to the SSB lump sum. Eligibility for VSI or SSB required more than six years and fewer than 20 years of active service. The services offered these pays selectively.

The Army

> offered VSI/SSB to enlisted members with more than 9 years of service in skills (defined by occupation and grade) that were overfilled. By and large, the main Army eligibility criteria were based on an individual's rank, YOS, and occupation. However, through the judicious choice of these variables, the Army was able to offer the benefit not only to overfilled occupations but also, implicitly, to some marginal performers. (Asch and Warner, 2001, p. 6)

Marginal performance was signaled by a long time in grade or a lower grade than typical for the total YOS. Empirical analysis found that the Army's implementation of VSI and SSB increased exits by 15.7 percent among eligible low-quality personnel and about 11 percent among eligible high-quality personnel. High-quality personnel were high school diploma graduates scoring in the upper half of the Armed Forces Qualification Test score distribution. About 40 percent of the Army personnel in Asch and Warner's data were high quality, and about 60 percent have been high quality in recent years (Hosek, Asch, and Mattock, 2012). At 40 percent high quality, the overall exit rate was 13.8 percent, and 32 percent of VSI- or SSB-induced leavers were high quality.[3] At 60 percent high quality, the overall exit rate would be 12.9 percent, assuming the same respective take rates, and 52 percent of VSI- and SSB-induced leavers would be high quality. The Asch and Warner study does not have comparable results for Army officers.

Beland and College, 1992, uses data on Army and Air Force officer and enlisted personnel to examine whether those who accepted the VSI or SSB offer were lower quality. The authors found that lower-quality personnel generally did accept the offer but that many who accepted would have left in any case. Mehay and Hogan, 1998, studies how separations changed as a result of the VSI and SSB program using data on Navy and Air Force enlisted personnel. The Navy differed from the Air Force in how it implemented the program and, specifically, unlike

[3] Percentage of leavers who were high quality among those who were induced to leave by VSI or SSB is

$$\frac{40 \times 0.11}{60 \times 0.157 + 40 \times 0.11} = 31.8.$$

With an Army of 60 percent high quality, the percentage is

$$\frac{60 \times 0.11}{40 \times 0.157 + 60 \times 0.11} = 51.2.$$

the Navy, the Air Force offer included an explicit threat that anyone who received but did not accept the offer would likely be involuntarily separated. Mehay and Hogan found a modest effect of the VSI and SSB program on Navy separations and, not surprisingly, a larger effect on Air Force separations. As explained in Asch and Warner, the Army policy was somewhere in between the Air Force's policy and the Navy's policy. That is, Army personnel were not explicitly told that anyone who received but did not accept the offer would be involuntarily separated, but personnel might have perceived that such a threat existed.

TERA pay was payable to members with more than 15 but fewer than 20 years of active service. TERA was computed like a retirement benefit: 0.025 times basic pay times YOS. This amount was multiplied by a "reduction factor based on the number of months the retiree is short of 20 years" (Defense Finance and Accounting Service, 2015).

Table 2.1 illustrates ISP, SSB, and VSI amounts for enlisted members and officers at typical grade/YOS combinations. The entries are based on the 2013 basic-pay table. The amount of VSI *to the service member* is shown as the present value (PV) of the annuity stream using personal discount rates of 10 percent for enlisted and 6 percent for officers. The table also shows the cost of VSI *to the government*, which is the amount the government must obligate in the current period to pay the VSI annuity given the cost of capital to the government.[4] Because of the difference between the personal discount rate and the cost of capital to the government, the value of VSI to the member is less than its cost to the government. ISP and SSB are lump sums, and their value to the member equals the cost to the government. ISP is the least generous of the three types of payment, SSB is intermediate, and VSI is the largest. Chapter Five further discusses ISP in comparison with VSP.

Voluntary Separation Pay

Like ISP, VSI, and SSB, VSP would be paid upon separation. VSP is not calculated from a given formula but determined through modeling and optimization described in this report. VSP is that value of separation pay sufficient to induce the desired number of personnel to leave voluntarily in accord with drawdown objectives. The objectives are to achieve a given decrease in personnel in a given period of time and to leave the resulting force as close to steady state as possible, avoiding management issues caused by over- or underdrawing the force. VSP is offered between seven and 18 YOS,[5] and its amount varies by YOS. VSP can be targeted, e.g., by rank, occupation, YOS, time in grade, and average speed of promotion (rank relative to YOS).

Unlike ISP, VSI, and SSB, VSP is designed to separate personnel voluntarily. ISP is a payment to personnel separated involuntarily; ISP does not need to be high enough to induce voluntary separation. We find that, in larger drawdowns, ISP is insufficient compensation for the

---

[4] We use the average real government interest rate for 2004–2013 for a 20-year maturity Treasury bond, which is 2.54 percent (Office of Management and Budget [OMB], 1992).

[5] Early- and late-career personnel qualify for other benefits when they separate. Early-career personnel are often eligible for Post-9/11 GI Bill (Title V of Pub. L. 110-252, 2008) educational benefits. Also, some early leavers have discovered that they are not a good fit with the military and would prefer to leave. Late in their careers, members with 20 or more YOS qualify for immediate retirement benefits and health benefits. These benefits are substantial and compensate for much of the loss from serving somewhat fewer years than expected. Overall, these other benefits seem likely to address concern about "breaking faith" for these groups.

Table 2.1
Illustrative Amounts of Involuntary Separation Pay, Voluntary Separation Incentive, and Special Separation Bonus, in Thousands of 2013 Dollars, Rounded

				VSI	
Grade	**YOS**	**ISP**	**SSB**	**Value to Member**	**Cost to Government**
Enlisted					
E-5	8	26	39	54	82
E-5	10	35	52	80	133
E-5	12	44	66	107	194
E-6	14	59	88	149	292
E-6	16	68	102	178	372
E-7	18	90	135	238	530
Officer					
O-3	8	51	77	132	163
O-3	10	68	101	200	259
O-4	12	94	141	306	416
O-4	14	115	173	403	572
O-4	16	136	204	502	741
O-5	18	171	256	655	1,006

loss of one's military career (Chapter Five). Further, VSI or SSB might be too small or too large to meet a specific drawdown goal; their formulaic approach, like that of ISP, is not designed to meet a specific goal.

VSP is more generous than ISP, and implementing it can be expected to cost more. However, ISP and the involuntary separation policies that accompany it might create negative externalities, e.g., off-budget costs, such as lower morale or higher effort to manage the personnel force structure that can result from involuntary mechanisms, inadequate levels of compensation, or separation pay levels that are incorrect with respect to achieving a given drawdown.

CHAPTER THREE

An Overview of the Approach

The DRM is well suited to the analysis of incentive pays. Recent applications of the model include analyses for the ninth, tenth, and 11th Quadrennial Reviews of Military Compensation; analyses of special and incentive pays for the Air Force; and analyses of military retirement reform proposals. Model capability has steadily increased with faster estimation and simulation programs and refined costing, and the model now can predict retention and cost effects in the steady state and transition to the steady state.

The DRM is a stochastic dynamic programming model of individual decisionmaking over the life cycle. The model's parameters are estimated using individual, longitudinal retention data from administrative data files. Starting at the beginning of AC service, the individual makes a stay/leave decision each year. Those who leave the AC take a civilian job and simultaneously decide whether to participate in the reserve component (RC). The decision of whether to participate in the RC is made each year, and the person can move into or out of the RC period by period. Interestingly, including reserve participation improves the precision of the empirical estimates of parameters governing active retention.[1]

The approach is documented in RAND reports, such as Mattock, Hosek, and Asch, 2012, a technical report prepared for the 11th quadrennial review; and Asch, Hosek, and Mattock, 2013, a report prepared for the Office of the Assistant Secretary of Defense for Manpower and Reserve Affairs. This work was further extended to consider the effect that policy changes can have on force size and shape over time during the transition to a new steady state in Asch, Mattock, and Hosek, 2013.[2] The appendix has a technical description of the model.

The present study extends the DRM to include a routine to find the optimal amount of separation pay to induce members in a particular cohort—that is, members at a particular YOS at the time of policy implementation—to separate voluntarily from the military to meet an end-strength goal for that cohort in a given period of time. This technique finds separation incentive pays that decrease any YOS cohorts by an arbitrary amount and bring the force overall to a size and shape that supports the new desired steady state. It can scale the force overall or scale selected occupations to maintain the experience profile while decreasing the strength to a desired level.

[1] The DRM has several limitations. The model assumes that military pay, promotion policy, and civilian pay are time-stationary, and it excludes demographic factors, such as gender, marital status, and spouse employment. It also excludes health status and health care benefits, and we do not explicitly model deployment or deployment-related pay. That said, the estimated models fit the observed data well for the both the AC and the RC.

[2] Goldberg, 2001, provides an extensive discussion of the history of retention models. Gotz, 1990, provides a detailed discussion of the advantages of the DRM approach relative to other approaches that have been used to assess the effects that compensation proposals can have on retention.

In the optimization routine, we want the new force to be near its steady state as soon as possible and want to avoid over- or undershooting the decrease of a cohort relative to its steady state. Overshooting would cause a shortage relative to the steady state, and the shortage would move through the force over time. Further, if the drawdown is not balanced by YOS and some cohorts are not decreased to a steady-state level, the experience mix could not be in steady state for many years.

The DRM is a behaviorally rich model of retention, its estimated parameters are statistically significant, and its predicted retention fits the data well. It is theoretically and empirically well grounded for determining the level of separation pay needed to meet drawdown targets without over- or undershooting the outflow of personnel. The DRM employs certain assumptions. Individuals are assumed to be forward-looking and able to assess the value of remaining in the AC for another year, which, in turn, depends on the value of remaining in the year after that and so on. Individuals can assess the value of leaving the AC to work as a civilian and participate in the RC, or not. Individuals make rational choices, which means that they select the best alternative given the information available to them each period. If there were changes to the structure of decisionmaking (e.g., changes not present in the retention data used to estimate the model), the model's predictions might not be good forecasts of future behavior if the separation incentive were implemented. If the model's specifications were inadequate (e.g., if it should have included additional variables), the predictions could be inaccurate. However, the model fits the data well. Finally, if an optimized VSP were implemented, there could be other conditions that affect retention. The model's predictions are a starting point for setting VSP, but actual implementation calls for flexibility to adjust VSP in view of prevailing conditions. This point also applies to implementing other drawdown approaches.

To describe our approach, consider VSP offered for one year only and not anticipated. We reasoned that, for a drawdown of x percent, a new force would be in a steady state if each drawn-down cohort had x percent fewer members at 19 YOS. For each cohort, we asked what amount of VSP would result in the cohort having x percent fewer members at YOS 19. We focus on 19 YOS because the distribution of taste for military service differs across the cohorts in the military as a result of selective retention; members with higher taste for military service are more likely to remain in service, and our research has shown mean taste to increase as YOS increase.[3] As a result, if we decreased cohorts by x percent at their YOS at policy implementation, cumulative retention to YOS 19 would differ by cohort, and the drawn-down force would not be in steady state. By comparison, using YOS 19 as the basis for computations controlled for differences in taste by cohort. Also, YOS 19 is the year before a member vests in the military retirement system. AC members with 20 YOS can retire with an immediate annuity. These benefits are themselves a form of separation pay that induces members to leave.

Our approach provides the same, steady-state cumulative retention as of YOS 19 for each current cohort being decreased in size by VSP. Also, although the steady-state level is not reached in the year VSP is offered, the retention decrease for each cohort in that year comes

[3] In the model, *taste* refers to a member-specific fixed effect, that is, a term that is constant over time. The model allows taste to differ from member to member, and estimation of the model identifies the taste distribution across members, i.e., the mean and variance of taste for active duty, the mean and variance of taste for duty in a service's RCs, and the correlation between active and reserve tastes. An individual's taste for service in an AC, for example, is the combined, net effect of the individual's preference for service, value placed on the amenities and disamenities of serving, and any persistent, unobserved differences between the individual's expected military earnings and earnings if a civilian (e.g., earning differences not captured in the average military and civilian pay lines used in estimating the model).

close to the steady-state decrease. That is, almost all of the decrease to the new steady state is obtained in the calendar year VSP is offered. The small remaining percentage is obtained by YOS 19. As a result, most of the cost savings occur right away, and the small difference from steady state in that year is justified by the objective of avoiding future imbalance. The imbalance would occur because, although a cut to steady-state level in the year VSP is offered would at first look optimal, retention in future years of those retained would differ from steady-state retention because of taste differences. Again, we eliminate these imbalances by computing a cohort's retention to YOS 19 and then working backward to find the VSP needed to generate the same cumulative retention to YOS 19 across all the drawn-down cohorts.

We considered multiyear windows for the drawdown, e.g., when members had two or three years to decide whether to accept VSP. A given percentage reduction at 19 YOS remained the goal, and we modified the individual's value of leaving the AC in the DRM to make the separation incentive payment available for a second or third year. As will be shown, VSP can be less when the window is wider. Intuitively, the chance that a given incentive will be acceptable is higher for two "draws" than one, so the same chance can be maintained by a lesser amount. Yet, because this allows individuals to strategically delay their departure, retention and personnel costs decrease more slowly.

Another aspect of policy design was whether to make an early announcement of the VSP policy. If the policy is announced a year or two before implementation, members can anticipate it. We find that leads to an increase in retention prior to implementation, which, in turn, requires higher VSP to achieve the same goal.

Table 4.3
Present Value of the Net Decrease in Personnel Costs over Five, Ten, and 30 Years, by Scenario, for Officer and Enlisted Personnel Together, in Billions of 2013 Dollars

		PV Net Decrease		
Scenario	**Initial Cost of VSP**	**Over 5 Years**	**Over 10 Years**	**Over 30 Years**
1-year window	2.0	2.9	7.4	10.7
2-year window	2.6	2.5	7.0	10.3
3-year window	1.7	2.1	6.6	9.9
Draw down all to 90%	2.0	2.9	7.4	10.7
Draw down half to 80%	2.4	2.5	7.0	10.4
Draw down one-quarter to 60%	3.0	1.8	6.4	9.9

NOTE: The change in personnel costs is only for personnel leaving in the years covered by the VSP, YOS 7 through 18. Discounting is done in accordance with OMB Circular A-94 (OMB, 1992), using real discount rates provided in Appendix C as updated in December 2012.

there have been more years of cost avoidance for the separated personnel. Also, personnel cost decreases continue to mount into the future. The PVs show, however, that three-fourths of the decreases have been realized by ten years from policy implementation.

CHAPTER FIVE

Concluding Thoughts

We review the main findings and discuss VSP with respect to ISP, VSI, and SSB.

Key Findings

- VSP can draw down the force rapidly without creating an imbalance in the experience mix of personnel (a bathtub).
- A higher VSP is need to make a deeper cut in force size, but net personnel costs decrease even so.
- A lower VSP offered for multiple years can achieve the same end-strength goal as a VSP offered for one year, but net decreases in personnel costs are lower because more personnel remain on hand while the VSP is offered.
- If members anticipate being offered a VSP, a higher VSP is required to reach a given target.
- A 10-percent cut in the force would produce a net decrease in PV of AC Army personnel costs of $6.4 billion to $7.4 billion over the first ten years.
- The cost of implementing VSPs is $1.7 billion to $3 billion for a 10-percent cut in the force, depending on whether the VSP is implemented for a window of one, two, or three years and the cut is across the board or concentrated on a portion of the force.
- VSPs can be targeted to particular groups, e.g., poorer performers or personnel in over-strength occupations, and thereby induce separations from those groups.

Comparing Voluntary Separation Pay with Involuntary Separation Pay, Voluntary Separation Incentive, and Special Separation Bonus

Targeting

VSP, ISP, VSI, and SSB can be targeted. A service can set eligibility criteria to focus on lower-performing members, for instance. This can be done along with actions that do not have a budget cost: lower accessions, early outs, tighter physical standards, tighter time-in-grade standards, tighter promotion selection standards, and longer time to promotion (because of a decrease in force structure and spaces). ISP, VSI, and SSB are paid by formulas depending on basic pay and YOS, while the DRM calculates VSP. The model has not yet been estimated for specific groups that might be targeted, but it could be.

Cost, Uncompensated Value, and Rent

The cost of implementing ISP, VSI, SSB, and VSP will depend on where the force is cut and how deeply. Because ISP, VSI, and SSB are computed by a preset formula, their budget cost is simply the number cut in each basic pay/YOS cell times the ISP, VSI, or SSB for that cell. The cost of VSP at a given YOS depends on the number of individuals opting to leave times the VSP for that YOS, which is a function of the depth of the cut. In contrast, ISP, VSI, and SSB amounts are independent of the size of the reduction.

Tables 5.1 and 5.2 illustrate the per-person cost of ISP and VSP for enlisted members and officers. (VSI and SSB are no longer available.) The VSP amounts are for a one-year implementation not announced ahead of time. ISP and VSP for a 10-percent enlisted drawdown are nearly equal except at YOS 16 and 18. Apart from those years, there is little difference between the costs of these two approaches. For drawdowns greater than 10 percent, ISP remains the same, but VSP increases with the depth of the drawdown. For officers, VSP exceeds ISP for a 10-percent drawdown, and VSP increases for higher drawdowns, while ISP remains at the

Table 5.1
Involuntary and Voluntary Separation Pay Payments, by Extent of Drawdown, Enlisted, in Thousands of 2013 Dollars

Percentage Decrease	8 YOS		10 YOS		12 YOS		14 YOS		16 YOS		18 YOS	
	ISP	VSP	ISP	VSP	ISP	VSP	ISP	VSP	ISP	VSP	ISP	VSP
10	26	25	35	30	44	41	59	59	68	88	90	126
15	26	34	35	40	44	53	59	73	68	103	90	142
20	26	43	35	50	44	63	59	84	68	116	90	155
25	26	50	35	58	44	72	59	94	68	126	90	166
30	26	57	35	65	44	81	59	103	68	135	90	175
35	26	64	35	73	44	89	59	111	68	144	90	183
40	26	71	35	80	44	96	59	119	68	152	90	191

Table 5.2
Involuntary and Voluntary Separation Pay Payments, by Extent of Drawdown, Officers, in Thousands of 2013 Dollars

Percentage Decrease	8 YOS		10 YOS		12 YOS		14 YOS		16 YOS		18 YOS	
	ISP	VSP	ISP	VSP	ISP	VSP	ISP	VSP	ISP	VSP	ISP	VSP
5	51	79	68	98	94	133	115	197	136	285	171	368
10	51	132	68	156	94	200	115	273	136	367	171	452
15	51	173	68	201	94	248	115	325	136	421	171	507
20	51	209	68	238	94	288	115	366	136	463	171	548
25	51	241	68	272	94	323	115	402	136	498	171	583
30	51	271	68	303	94	355	115	434	136	530	171	614
35	51	299	68	332	94	384	115	463	136	558	171	642
40	51	327	68	360	94	412	115	491	136	586	171	669

same level. For example, at 12 YOS, ISP for enlisted is about $44,000, and VSP is $41,000 for a 10-percent drawdown and $96,000 for a 40-percent drawdown. ISP for officers at 12 YOS is $94,000, and VSP is $133,000 for a 5-percent drawdown, $200,000 for a 10-percent drawdown, and $412,000 for a 40-percent drawdown. Also, for a drawdown of a given size, the required VSP increases nonlinearly with YOS. The VSP for a 10-percent enlisted drawdown is $44,000 at 12 YOS, $88,000 at 16 YOS, and $126,000 at 18 YOS. The respective VSPs for officers are $200,000, $367,000, and $452,000.

But as noted in Chapter Four, the roles of ISP and VSP are fundamentally different. ISP is a payment to members selected by and involuntarily separated by the service. VSP is a payment to members who self-select and voluntarily choose to separate from the service. The VSP amounts computed by the DRM are for self-selection and voluntary separation and are not intended to indicate the amount of VSP required to cut a given number of people from a population of members designated for *involuntary* separation.

Under voluntary separation, VSP induces the *n*th member to separate to meet a drawdown of *n* members. This amount is higher than needed to separate the $n - 1$ others, so they receive rent (are paid more than they would be willing to accept to leave). VSP is an appropriate tool when well-performing personnel must be cut because of a general drawdown. Under involuntary separation, ISP can often be less than a member would be willing to accept to leave voluntarily, and, if so, the member suffers a loss. The service sets the criteria for involuntary separation, and there could be negative externalities if members think the criteria are unfair (do not have the same bearing across occupations, experience, or demographic groups), inaccurate or imprecise (poorly discriminate among personnel based on their performance), or too deep (the criteria are perceived to be valid for identifying the very lowest performers but invalid when affecting a larger portion of the force). Also, many of those selected for involuntary separation might feel that ISP is not fair compensation. Equally important, those not selected might disapprove of the extensive use of an involuntary mechanism in a volunteer force and think that it breaks faith with them.

In summary, involuntary separations seem appropriate for relatively small cuts of perhaps 10 percent or less and when criteria are meaningful to the service and perceived as valid by service members. Deeper involuntary separations and less valid criteria, however, could create negative externalities among members, a hidden cost. Also, if involuntary separations are not carefully done, they could create bathtubs. Voluntary separations under VSP provide a balanced experience mix and avoid negative externalities, but they are more costly, especially for deeper cuts, and deprive the service of controlling who separates apart from setting the criteria to define the eligible population.

APPENDIX

The Dynamic Retention Model

The following material comes from Asch, Hosek, and Mattock, 2013, pp. 81–88, and is reproduced here for the convenience of the reader. Note that we make one correction to the final paragraph, which refers to the BHHH method and cites Berndt et al., 1974. BHHH (better known as the outer-product-of-gradients method) is a means of approximating the variance–covariance matrix as a precursor to computing standard errors and has nothing to do with optimization. The method that was used was used for optimization was Broyden–Fletcher–Goldfarb–Shanno (BFGS) (Broyden, 1970). The corrected text would read as follows:

> We use the Broyden–Fletcher–Goldfarb–Shanno (BFGS) hill-climbing algorithm to optimize the likelihood function (Broyden, 1970). We compute standard errors by numerically differentiating the likelihood function, evaluated at parameter estimates to produce a matrix of second derivatives, or Hessian matrix. The standard errors are the square root of the absolute value of the diagonal of the inverse of the Hessian.

APPENDIX B

The Active/Reserve Dynamic Retention Model

As discussed in Chapter Two, the model is a stochastic dynamic programming model of active retention and reserve participation at the individual level. The model is a theoretical basis for describing behavior where the individual is assumed to be rational and forward looking. In dynamic programming models, the current state depends on history, i.e., on the sequence of past states, and the decision taken in each period. The model is stochastic in the sense that in each period random factors enter the decision. There is a realization of the random factors in each period once it is reached. The realizations in future periods are not known in the current period, but the individual is assumed to know the distribution from which the random factors are drawn and can use this knowledge in developing an expected value of the future consequences of the current-period choice. The following expressions give the structure of the model. Again, the model is defined at the individual level, however, the individual subscript is suppressed for brevity.

$$Y_{jkt}(s_t, \varepsilon_{kt}; \gamma) = w_{kt} + \gamma_k + \beta \, \text{Emax}(Y_{ka}(s_{t+1}, \varepsilon_{at+1}; \gamma), Y_{kr}(s_{t+1}, \varepsilon_{rt+1}; \gamma), Y_{kc}(s_{t+1}, \varepsilon_{ct+1}; \gamma)) + \varepsilon_{kt} \tag{B.1}$$

Y_{jk}= value function for transition from j to k, j, $k \in$ {active, reserve, civilian}
$s_t = (ay_t, ry_t, t)$ where ay = active years, ry = reserve years, t = total years
w_{kt}= current pay in k at t

$$\gamma_k = \begin{cases} \gamma_a & \text{monetary value of preference for serving in AC} \\ \gamma_r & \text{monetary value of preference for serving in RC} \\ 0 & \text{preference for civilian job} \end{cases}$$

β = personal discount factor
Emax = expected value of the maximum
ε_{kt} = random shock in k at t.

The model is generally structured to allow movement between active, reserve, and civilian statuses, but in applying the model we do not permit an individual who leaves an active component to reenter. This decreases the state space, which facilitates the estimation of the model, and reflects the fact that reentry is relatively rare. The value function Y_{jkt} subscripts indicate current status j, a status k that the individual can enter next period from j, and the time period. The value function is additively separable in the current pay in k, the monetary value of taste in k, the present value of being able to choose the best alternative in the following period given k in the current period, and the random shock in k in the current period.

The state, s_t, is defined in terms of active years, reserve years, and total years accumulated as of period t. Current pay depends on the state. Active pay is average annual regular military compensation given the number of active years and reserve years. Reserve pay is a fraction of reserve annual regular military compensation under the assumption that a reservist accumulates 75 points in a year and receives 75/360 of annual reserve RMC (Appendix A). Pay as a reservist, i.e., in the reserve status, is the sum of reserve pay and civilian pay, an approach that assumes reservists are also employed in civilian jobs. Civilian pay depends on total years. In addition, pay in the civilian status includes the present discounted value of the active or reserve military retirement benefit payment if the individual is eligible to receive it.

The term γ_k is the monetary value of the individual's preference relative to the civilian sector; i.e., γ_a for active service and γ_r for reserve service. The personal discount factor, β, is defined as 1/1 + r, where r is the personal discount rate. The operator Emax gives the expected value of the *maximum* of the value functions in the next period. The Emax expression reflects the fact that the individual can reoptimize in the next period once the random shocks in that period have been realized. The current period assessment of the value of the best choice tomorrow is the expected value of the maximum of tomorrow's choices. The term ε_{kt} is the random shock in status k in period t.

The model is structured as a Markov process. In the next period there is a chance that any allowable status can be entered. Further, because the state is assumed to capture all relevant information from the individual's history and the random shocks are uncorrelated, it is possible to partition the expected value of the maximum given the current state. Using this insight, the model also can be written:

$$Y_{jkt}(s_t) = w_{kt} + \gamma_k + \beta \sum_m \pi_{km}(s_{t+1} \mid s_t) Y_{km}(s_{t+1}) + \varepsilon_{kt}$$

$$\pi_{km}(s_{t+1} \mid s_t) = \text{probability alternative } m \text{ is } \max(Y_{km}(s_{t+1})), m \in \{a,r,c\}. \quad \text{(B.2)}$$

The model assumes that a reservist holds a civilian job. This is a simplifying assumption because some reservists are full-time students, unemployed, or out of the labor force, but the idea is that participation in the reserves is concurrent with another main activity, a job. Therefore, a civilian job shock will be present in both the civilian and reserve statuses.

To allow for error correlation between the reserve and civilian alternatives, we use a nested logit form where these alternatives represent one nest and the active alternative is the other nest. The choice is between the active alternative and the better alternative in the reserve/civilian nest, i.e., the maximum of the reserve alternative and the civilian alternative. To shorten notation, we rewrite Equation B.1 as $Y_{kj}(s_t)=V_j+\varepsilon_j$, where V_j represents the non-stochastic terms on the right side, and the other arguments and time subscript have been omitted. Adapting the treatment of Ben-Akiva and Lerman (1985), we develop the nested logit specification from the following expressions:

$$V_a+\varepsilon_a$$
$$\max[V_r+\omega_r, V_c+\omega_c] + \upsilon_{rc}. \tag{B.3}$$

The first expression in Equation B.3 corresponds to the active alternative, and the second expression corresponds to the reserve/civilian nest alternative. The active alternative can be thought of as a nest with a single element. The nested logit model assumes that ε_a has the same distribution as the sum of the errors in the second expression, so we need to ensure that this requirement is met. Also, we assume that all errors are generated from extreme value distributions. When the errors have the same extreme value distribution, and in particular have the same variance, then the choice between the nests has the logit form. Train (2003) provides a proof that when alternatives have identically distributed, independent extreme-value errors, the probability that a particular alternative is the maximum has a logit form. Ben-Akiva and Lerman (1985) show that the nested logit model can be written as a choice between alternatives, each of which is the maximum choice from its nest. As we show for our model, the errors of these maximum choices can be constructed to have the same variance; hence, Train's proof applies.

The extreme value distribution $EV[a,b]$ has the form $e^{-e^{(a-x)/b}}$ with mean $a+b\gamma$ and variance $\pi^2b^2/6$, where γ is Euler's Gamma (≈0.577), a is the location parameter, and b is the scale parameter. The variance is proportional to the square of the scale parameter, and we use the fact that equal scale parameters imply equal variances. Let ω_r and ω_c in Equation B.3 be within-nest errors drawn from an extreme-value distribution $EV[0,\lambda]$ and let υ_{rc} be the nest-specific error for the reserve/civilian nest in Equation B.3, distributed as $EV[0,\tau]$. In other words, υ_{rc} can be thought of as a shock that affects both the reserve and the civilian alternatives, whereas ω_r and ω_c affect each alternative separately.

It is known that $\max[V_r+\omega_r, V_c+\omega_c]$ also follows an extreme-value distribution with the same scale as for ω_r and ω_c but a different mean, namely, $EV\left[\lambda \ln(e^{V_r/\lambda}+e^{V_c/\lambda}),\lambda\right]$. Notice that this mean is positive, assuming V_r and V_c are positive, whereas the distribution for ω_r and ω_c has a zero mean. Intuitively, the expected value of being able to choose the larger of two random draws, each with zero mean, is greater than zero. We rewrite the second expression in Equation B.3 as follows:

$$\lambda \ln\left(e^{V_r/\lambda} + e^{V_c/\lambda}\right) + \omega'_{rc} + \nu_{rc}, \text{ where}$$
$$\omega'_{rc} = \max\left[V_r + \omega_r, V_c + \omega_c\right] - \lambda \ln\left(e^{V_r/\lambda} + e^{V_c/\lambda}\right)$$
$$\omega'_{rc} \sim EV[0,\lambda]. \tag{B.4}$$

Define $\varepsilon_{rc} = \omega'_{rc} + \upsilon_{rc}$. It is the sum of two independent, differently distributed extreme-value variables. The error ω'_{rc} is the single error associated with taking the maximum of $V_r + \omega_r$ and $V_c + \omega_c$, and the definition of ω'_{rc} ensures that its mean is zero. Further, υ_{rc} is the single error at the nest level. The distributions of ω'_{rc} and υ_{rc} have the same location parameter (zero), but different scale parameters. In general, the variance of the sum of two independent random variables is the sum of the variances, so the variance of $\varepsilon_{rc} = \omega'_{rc} + \upsilon_{rc}$ is $\left(\pi^2\left(\lambda^2 + \tau^2\right)\right)/6$, implying a scale parameter for the R/C nest of $\sqrt{\lambda^2 + \tau^2}$. It follows that $\varepsilon_{rc} \sim EV\left[0, \sqrt{\lambda^2 + \tau^2}\right]$. We also want ε_a to have the same distribution (i.e., the same location and scale parameters), so we set $\varepsilon_a \sim EV\left[0, \sqrt{\lambda^2 + \tau^2}\right]$. For brevity, let $\kappa = \sqrt{\lambda^2 + \tau^2}$.

Drawing this together, the model may be written as follows:

$$V_a + \varepsilon_a$$
$$\lambda \ln\left(e^{V_r/\lambda} + e^{V_c/\lambda}\right) + \varepsilon_{rc}$$
$$\varepsilon_a, \varepsilon_{rc} \sim EV[0,\kappa]. \tag{B.5}$$

Assuming that the individual chooses the higher-valued alternative, this leads to a probability of choosing *active* that has the logit form, as Train (2003) showed:

$$\Pr(active) = \frac{e^{V_a/\kappa}}{e^{V_a/\kappa} + e^{\lambda \ln\left(e^{V_r/\lambda} + e^{V_c/\lambda}\right)/\kappa}}$$
$$= \frac{e^{V_a/\kappa}}{e^{V_a/\kappa} + \left(e^{V_r/\lambda} + e^{V_c/\lambda}\right)^{\lambda/\kappa}}. \tag{B.6}$$

The second line follows from the fact that $e^{b \ln a} = a^b$.

The within-nest error terms, ω, are distributed $EV[0,\lambda]$ and the "total" error terms, ε, are distributed $EV\left[0, \sqrt{\lambda^2 + \tau^2}\right]$.

Therefore, the fraction of the error variance accounted for by the within-nest, choice-specific, portion of the total error is

$$\frac{\lambda^2}{\tau^2 + \lambda^2}. \tag{B.7}$$

It follows that the fraction of the error variance attributable to the within-nest common shock is one minus this amount, or $\tau^2/\left(\tau^2 + \lambda^2\right)$.

As a thought experiment, we can think of the problem of selecting the best alternative from the nest as choosing between

$$\begin{gathered} V_r + \omega_r + \upsilon_{rc} \\ V_c + \omega_c + \upsilon_{rc} \end{gathered} \tag{B.8}$$

The correlation between these two total utilities (viewed by themselves before one has been chosen) is

$$\begin{aligned} \rho &= \frac{Cov[V_r + \omega_r + \nu_{rc}, V_c + \omega_c + \nu_{rc}]}{\sqrt{Var[V_r + \omega_r + \nu_{rc}]}\sqrt{Var[V_c + \omega_c + \nu_{rc}]}} \\ &= \frac{\tau^2}{\tau^2 + \lambda^2}. \end{aligned} \tag{B.9}$$

As shown in Equation B.9, a larger variance of the common shock results in a larger correlation between the reserve and civilian alternatives. Thus, the nested logit formulation succeeds in giving us a specification that allows the shocks to the reserve and civilian alternatives to be correlated, and the greater the common shock, the greater the correlation.

Applying the rule above for the distribution of the maximum of two values, we see that

$$\begin{gathered} \max\left[V_a + \varepsilon_a, \lambda \ln\left(e^{V_r/\lambda} + e^{V_c/\lambda}\right) + \varepsilon_{rc}\right] \sim EV\left[\kappa \ln\left(e^{V_a/\kappa} + e^{\lambda \ln\left(e^{V_r/\lambda} + e^{V_c/\lambda}\right)/\kappa}\right), \kappa\right] \\ EV\left[\kappa \ln\left(e^{V_a/\kappa} + e^{\lambda \ln\left(e^{V_r/\lambda} + e^{V_c/\lambda}\right)/\kappa}\right), \kappa\right] = EV\left[\kappa \ln\left(e^{V_a/\kappa} + \left(e^{V_r/\lambda} + e^{V_c/\lambda}\right)^{\lambda/\kappa}\right), \kappa\right]. \end{gathered} \tag{B.10}$$

As before, the second line follows from $e^{b \ln a} = a^b$.

Applying the formula for the mean of an extreme-value distribution to Equation B.10, the expected value of the maximum of the two alternatives (active versus the maximum of reserve/civilian), is

$$\kappa\left(\gamma + \ln\left[e^{V_a/\kappa} + \left(e^{V_r/\lambda} + e^{V_c/\lambda}\right)^{\lambda/\kappa}\right]\right). \tag{B.11}$$

Further, given that active is not an option, the expected value of the maximum of the two alternatives (reserve and civilian) is

$$\begin{aligned} &\kappa\left(\gamma + \ln\left[\left(e^{V_r/\lambda} + e^{V_c/\lambda}\right)^{\lambda/\kappa}\right]\right) \\ &= \kappa\gamma + \lambda \ln\left[e^{V_r/\lambda} + e^{V_c/\lambda}\right]. \end{aligned} \tag{B.12}$$

The first line of Equation B.12 does not contain the term $e^{V_a/\kappa}$ because the constraint that the individual cannot reenter the active component means, in effect, that V_a is set to negative infinity, and $e^{-\infty} = 0$. The second line simplifies the log expression.

The expected value of the maximum of a set of choices is referred to as the surplus function, and the surplus function can be used to derive choice probabilities. The Williams-Daly-Zachary Theorem (see McFadden, 1981) states that the probability of choosing a given alternative equals the partial derivative of the surplus function with respect to the value of the alternative. Thus, the probability of choosing to remain in an active component is as follows:

$$\Pr(active) = \frac{\partial \kappa\left(\gamma + \ln\left[e^{V_a/\kappa} + \left(e^{V_r/\lambda} + e^{V_c/\lambda}\right)^{\lambda/\kappa}\right]\right)}{\partial V_a} = \frac{e^{V_a/\kappa}}{e^{V_a/\kappa} + \left(e^{V_r/\lambda} + e^{V_c/\lambda}\right)^{\lambda/\kappa}}. \quad \text{(B.13)}$$

This is the same as that shown in Equation B.6, which replicated the usual logit specification. To emphasize the meaning of Equation B.13, we restate it as

$$PR\left(V_a + \varepsilon_a > \lambda \ln\left(e^{V_r/\lambda} + e^{V_c/\lambda}\right) + \varepsilon_{rc}\right) = \frac{e^{V_a/\kappa}}{e^{V_a/\kappa} + \left(e^{V_r/\lambda} + e^{V_c/\lambda}\right)^{\lambda/\kappa}}.$$

By the same approach, the probabilities of choosing reserve and civilian are

$$\Pr(reserve) = \frac{\left(e^{V_r/\lambda} + e^{V_c/\lambda}\right)^{\lambda/\kappa}}{e^{V_a/\kappa} + \left(e^{V_r/\lambda} + e^{V_c/\lambda}\right)^{\lambda/\kappa}} \frac{e^{V_r/\lambda}}{e^{V_r/\lambda} + e^{V_c/\lambda}} \quad \text{(B.14)}$$

$$\Pr(civilian) = \frac{\left(e^{V_r/\lambda} + e^{V_c/\lambda}\right)^{\lambda/\kappa}}{e^{V_a/\kappa} + \left(e^{V_r/\lambda} + e^{V_c/\lambda}\right)^{\lambda/\kappa}} \frac{e^{V_c/\lambda}}{e^{V_r/\lambda} + e^{V_c/\lambda}}.$$

Given that the individual has left the active component and cannot reenter it, the probabilities of choosing reserve or civilian are, respectively,

$$\Pr(reserve \mid inactive) = \frac{\partial\left(\kappa\gamma + \lambda Log\left[e^{V_r/\lambda} + e^{V_c/\lambda}\right]\right)}{\partial V_r} = \frac{e^{V_r/\lambda}}{e^{V_r/\lambda} + e^{V_c/\lambda}} \quad \text{(B.15)}$$

$$\Pr(civilian \mid inactive) = \frac{\partial\left(\kappa\gamma + \lambda Log\left[e^{V_r/\lambda} + e^{V_c/\lambda}\right]\right)}{\partial V_c} = \frac{e^{V_c/\lambda}}{e^{V_r/\lambda} + e^{V_c/\lambda}}.$$

A comparison of Equations B.14 and B.15 shows that the probability of choosing to be a reservist equals the probability of choosing the reserve/civilian nest multiplied by the probability of choosing reserve, given that the individual is in the nest. A similar statement holds for the probability of choosing to be a civilian.

The model provides structure regarding the choice among alternatives. The choice is modeled to depend additively on current pay, taste for active or reserve service, current shock, and the expected value of rational choice among alternatives in the uncertain future, i.e., the expected value of the maximum. Further structure comes from assuming the individual knows the shock variances and so has the information needed to compute the expected value of the maximum. The individual also knows the civilian, reserve, and active pay lines. With this information, along with knowing the current state and shock draws, the individual can solve the problem and determine which alternative is best.

The position of the analyst is different. The analyst does not know the individual's tastes for active and reserve service, nor the current shocks, but is assumed to know their type of distribution. In particular, we assume that tastes follow a bivariate normal distribution and shocks follow generalized extreme value distributions, as assumed in the nested logit model. The bivariate normal distribution has five parameters: mean active taste, mean reserve taste, active taste variance, reserve taste variance, and active-reserve taste covariance or correlation. The extreme value distributions for the shocks have zero location parameters and non-zero scale parameters, hence non-zero variances.

In addition, we include the personal discount factor and switching cost parameters. The latter represent the cost of switching across states. There are switching costs for switching from active in the first two years of active service, switching from active in later years, and switching from civilian to reserve status. Estimation of the model involves finding the taste, shock, personal discount, and switching cost parameters that fit the data best, where the data consist of longitudinal observations on the individual's sequence of active, reserve, and civilian statuses over the work life.

Unlike the individual, who is assumed to have the information to make an explicit choice each period, the analyst does not have information about the individual's tastes or shocks. But the analyst can make use of the model and the functional form of the shock distributions to write an expression for the probability of a particular choice given the individual's state, and in this way can compose an expression for likelihood for the sequence of statuses over the individual's career. Still, this expression is conditional on the individual's tastes. Because these tastes are unknown to the analyst, they need to be integrated out of the expression, where the integration uses the assumption that the tastes have a bivariate normal distribution.

In estimation, the integration is done numerically. For each individual, a Halton sequence of 23 pairs of active and reserve seed tastes is drawn and then, using trial values of the taste distribution parameters, the Halton draws are translated as though they were drawn from the distribution of tastes given the trial values of the parameters. The translation is done via a Cholesky decomposition (Appendix C). For each resulting pair, the dynamic program is solved, giving values of the value functions at each decision point and hence values of the individual's career likelihood. The integration over tastes is accomplished by taking the average of the likelihoods over the 23 valuations.

The process of estimation tries different values of the parameters until the career likelihoods are maximized for the sample of service members used in the estimation. In many respects, this is standard maximum likelihood estimation, but it differs in two ways. First, the model has a specific structure for the value function, as mentioned above. Second, for each set of trial parameters, the dynamic programming problem must be re-solved for all periods for all 23 pairs of taste draws for each individual. Then, given the new solution of the model, the choice probabilities are updated, and the likelihood function is reevaluated to determine whether the fit has improved and in what direction the distribution parameters should be further changed in proceeding to the next iteration of estimation. Re-solving the dynamic program requires extensive computation. Estimating the Hessian matrix to determine the optimal direction of change is also computationally time consuming.

As the estimation algorithm iterates, we can think of the effect of changing the shock variances while holding constant the taste distribution parameters. An increase in a shock variance improves the fit if it does a better job of accounting for transitions from active to reserve, active to civilian, civilian to reserve, and reserve to civilian. That is, changing the variance affects all transitions, given any set of taste distribution parameters. Reasoning similarly, changing the mean active taste affects the fit of the active/active, active/reserve, and active/civilian transitions but not the other transitions. Changing the active taste variance helps account for dispersion in the transition probabilities from active duty. A change in the taste covariance affects the degree to which longer active careers are associated with longer reserve careers, e.g., higher transitions from civilian to reserve and lower transitions from reserve to civilians. Similar reasoning applies to the reserve taste mean and variance, with the implication that all the taste and shock parameters are identified. The personal discount rate is also estimable as it decreases future values relative to present values until the best fit is achieved. The switching cost parameters further improve the fit of the model.

We use the BHHH hill-climbing algorithm to optimize the likelihood function (Berndt et. al., 1974). We compute standard errors by numerically differentiating the likelihood function, evaluated at parameter estimates to produce a matrix of second derivatives, or Hessian matrix. The standard errors are the square root of the absolute value of the diagonal of the inverse of the Hessian.

References

Asch, Beth J., James Hosek, and Michael G. Mattock, *A Policy Analysis of Reserve Retirement Reform*, Santa Monica, Calif.: RAND Corporation, MG-378-OSD, 2013. As of November 24, 2015:
http://www.rand.org/pubs/monographs/MG378.html

Asch, Beth J., Michael G. Mattock, and James Hosek, *A New Tool for Assessing Workforce Management Policies over Time: Extending the Dynamic Retention Model*, Santa Monica, Calif.: RAND Corporation, RR-113-OSD, 2013. As of November 24, 2015:
http://www.rand.org/pubs/research_reports/RR113.html

Asch, Beth J., and John T. Warner, *An Examination of the Effects of Voluntary Separation Incentives*, Santa Monica, Calif.: RAND Corporation, MR-859-OSD, 2001. As of November 24, 2015:
http://www.rand.org/pubs/monograph_reports/MR859.html

Beland, Russell, and Craig College, "The Effects of Personnel Quality of Downsizing," Washington, D.C.: Office of the Secretary of Defense, Program Analysis and Evaluation, unpublished paper, 1992.

Ben-Akiva, Moshe E., and Steven R. Lerman, *Discrete Choice Analysis: Theory and Application to Travel Demand*, Cambridge, Mass.: MIT Press, 1985.

Berndt, Ernst K., Bronwyn H. Hall, Robert E. Hall, and Jerry A. Hausman, "Estimation and Inference in Nonlinear Structural Models," *Annals of Economic and Social Measurement*, Vol. 3/4, 1974, pp. 653–665.

Broyden, C. G., "The Convergence of a Class of Double-Rank Minimization Algorithms: 1. General Considerations," *IMA Journal of Applied Mathematics*, Vol. 6, No. 1, 1970, pp. 76–90.

Defense Finance and Accounting Service, "1993–2001 Temporary Early Retirement Authority," page updated March 31, 2015. As of January 24, 2013:
http://www.dfas.mil/retiredmilitary/plan/retirement-types/tera.html

Goldberg, Matthew S., *A Survey of Enlisted Retention: Models and Findings*, Alexandria, Va.: CNA Corporation, CRM D0004085.A2/Final, November 2001. As of November 24, 2015:
https://www.cna.org/CNA_files/PDF/D0004085.A2.pdf

Gotz, Glenn A., "Comment on 'The Dynamics of Job Separation: The Case of Federal Employees,'" *Journal of Applied Econometrics*, Vol. 5, No. 3, July–August 1990, pp. 263–268.

Hosek, James, Beth J. Asch, and Michael G. Mattock, *Should the Increase in Military Pay Be Slowed?* Santa Monica, Calif.: RAND Corporation, TR-1185-OSD, 2012. As of May 10, 2016:
http://www.rand.org/pubs/technical_reports/TR1185.html

Mattock, Michael G., James Hosek, and Beth J. Asch, *Reserve Participation and Cost Under a New Approach to Reserve Compensation*, Santa Monica, Calif.: RAND Corporation, MG-1153-OSD, 2012. As of November 24, 2015:
http://www.rand.org/pubs/monographs/MG1153.html

McFadden, Daniel L., "Econometric Models of Probabilistic Choice," in Charles F. Manski and Daniel L. McFadden, eds., *Structural Analysis of Discrete Data with Econometric Applications*, Cambridge, Mass.: MIT Press, 1981, pp. 198–272.

Mehay, Stephen L., and Paul F. Hogan, "The Effect of Separation Bonuses on Voluntary Quits: Evidence from the Military's Downsizing," *Southern Economic Journal*, Vol. 65, No. 1, July 1998, pp. 127–139.

Office of Management and Budget, *Guidelines and Discount Rates for Benefit–Cost Analysis of Federal Programs*, Washington, D.C., Circular A-94, October 29, 1992. As of November 29, 2015:
https://www.whitehouse.gov/omb/circulars_a094/

Office of the Under Secretary of Defense for Personnel and Readiness, *The 11th Quadrennial Review of Military Compensation: Main Report*, Washington, D.C., June 2012. As of November 24, 2015:
http://oai.dtic.mil/oai/oai?verb=getRecord&metadataPrefix=html&identifier=ADA563239

OMB—*See* Office of Management and Budget.

Public Law 110-252, Supplemental Appropriations Act of 2008, June 30, 2008. As of November 30, 2015:
http://www.gpo.gov/fdsys/pkg/PLAW-110publ252/content-detail.html

Public Law 112-81, National Defense Authorization Act for Fiscal Year 2012, December 31, 2011. As of November 30, 2015:
http://www.gpo.gov/fdsys/pkg/PLAW-112publ81/content-detail.html

Train, Kenneth, *Discrete Choice Methods with Simulation*, New York: Cambridge University Press, 2003.

Under Secretary of Defense for Personnel and Readiness, "Eligibility of Regular and Reserve Personnel for Separation Pay," Department of Defense Instruction 1332.29, June 20, 1991, incorporating change 2, September 20, 2011. As of November 30, 2015:
http://www.dtic.mil/whs/directives/corres/pdf/133229p.pdf

Warner, John T., and Saul Pleeter, "The Personal Discount Rate: Evidence from Military Downsizing Programs," *American Economic Review*, Vol. 91, No. 1, March 2001, pp. 31–51.

Office of the Secretary of Defense Cost Assessment and Program Evaluation requested RAND National Defense Research Institute help in developing an efficient means of decreasing force size. The researchers proposed the use of voluntary separation pay (VSP) and showed how it can be designed to meet drawdown goals within a certain time frame without over- or undershooting the goals. RAND's dynamic retention model determined the appropriate VSP levels by year of service to achieve drawdowns of alternative sizes for the active component Army, and the approach could be applied to other services. The analysis was done for enlisted personnel and officers and for the steady state and the transition to it.

The analysis suggests that VSPs can draw down the force rapidly without creating a hollow force or bathtub. Implementing VSPs requires an increase in outlays, but net decreases in personnel costs can still be realized even if selected military occupational specialties are cut deeply. The net decrease in personnel cost is less if a VSP is offered for multiple years, if members anticipate that a VSP will be offered in a future year, or if deeper cuts are made to portions of the force versus a shallower across-the-board cut. For a 10-percent cut in the force, net decreases range from $6.4 billion to $7.4 billion in 2013 dollars over the first ten years. The Army would initially require $1.7 billion to $3 billion to implement VSPs for such a cut, depending on how it is done.

www.rand.org

$24.50

ISBN-10 0-8330-9231-6
ISBN-13 978-0-8330-9231-1
52450
9 780833 092311

RR-545-OSD

Research Report

Implementation Actions for Improving Air Force Command and Control Through Enhanced Agile Combat Support Planning, Execution, Monitoring, and Control Processes

Kristin F. Lynch, John G. Drew, Robert S. Tripp, Daniel M. Romano, Jin Woo Yi, Amy L. Maletic

RAND Project AIR FORCE

Prepared for the United States Air Force
Approved for public release; distribution unlimited

The research described in this report was sponsored by the United States Air Force under Contract FA7014-06-C-0001. Further information may be obtained from the Strategic Planning Division, Directorate of Plans, Hq USAF.

Library of Congress Cataloging-in-Publication Data is available for this publication.

ISBN: 978-0-8330-8141-4

RAND OFFICES

SANTA MONICA, CA • WASHINGTON, DC

PITTSBURGH, PA • NEW ORLEANS, LA • JACKSON, MS

BOSTON, MA • CAMBRIDGE, UK • BRUSSELS, BE

www.rand.org

Preface

There has always been disparity between the availability of combat support resources and process performance and the capabilities needed to support military operations. Therefore, operational commanders, the authorities who prioritize and allocate scarce resources among operational commanders, and resource providers need to know how combat support enterprise constraints and alternative resource allocation decisions would impact planned and potential operations. They also need to know when agile combat support (ACS) process performance breaches the control parameters set for specific contingency operations. Currently, ACS planning, execution, monitoring, and control processes are not adequately defined and delineated in doctrine, guidance, and instructions. In addition, the tools, systems, training, and organizations needed to execute these ACS processes are lacking.

The focus of this analysis is on how enhanced ACS processes can be implemented and integrated into the Air Force and Joint command and control (C2) enterprise. Using the vision for enhanced C2 provided in the updated architecture developed as a companion piece to this analysis,[1] we identify and describe where shortfalls or major gaps exist between current ACS processes (the AS-IS) and the vision for integrating enhanced ACS processes into Air Force C2 (the TO-BE). We evaluate C2 nodes from the level of the President and Secretary of Defense (SECDEF) to the units and sources of supply. We also evaluate these nodes across the operational phases, from readiness preparation through planning, deployment, employment, sustainment, and reconstitution.

The research reported here was commissioned by the Deputy Chief of Staff for Logistics, Installations, and Mission Support (AF/A4/7) and the Vice Commander of the Air Force Materiel Command (AFMC/CV) and was conducted within the Resource Management Program of RAND Project AIR FORCE (PAF) as part of a project titled "Quantifying and Reducing Operational Risk."

This report will interest combatant commanders (CCDRs) and their staffs, component numbered Air Forces (C-NAFs) and their staffs, logisticians, planners, operators, and

---

[1] Kristin F. Lynch, John G. Drew, Robert S. Tripp, Daniel M. Romano, Jin Woo Yi, and Amy L. Maletic, *An Operational Architecture for Improving Air Force Command and Control Through Enhanced Agile Combat Support Planning, Execution, Monitoring, and Control Processes*, Santa Monica, Calif.: RAND Corporation, RR-261-AF, 2014. This report describes, in detail, a strategic- and operational-level C2 architecture integrating enhanced ACS processes.

employers of air and space C2 capabilities throughout the U.S. Department of Defense (DoD), especially those involved with C2 of forces during combat operations.

This document is one of a series of RAND publications that address combat support issues. Related publications include the following:

- Robert S. Tripp, Kristin F. Lynch, John G. Drew, and Robert DeFeo, *Improving Air Force Command and Control Through Enhanced Agile Combat Support Planning, Execution, Monitoring, and Control Processes*, Santa Monica, Calif.: RAND Corporation, MG-1070-AF, 2012. This monograph compares the current state of ACS planning, execution, monitoring, and controlling with the suggested implementation actions designed to address shortfalls identified in the 2002 PAF operational architecture. It further recommends implementation strategies to facilitate changes needed to improve Air Force C2 through enhanced ACS planning, execution, monitoring, and control processes.
- Kristin F. Lynch and William A. Williams, *Combat Support Execution Planning and Control: An Assessment of Initial Implementations in Air Force Exercises*, Santa Monica, Calif.: RAND Corporation, TR-356-AF, 2009. This report evaluates the progress the Air Force has made in implementing the TO-BE ACS operational architecture as observed during operational-level C2 warfighter exercises Terminal Fury 2004 and Austere Challenge 2004 and identifies areas that need to be strengthened. By monitoring ACS processes, such as how combat support requirements for force package options that were needed to achieve desired operational effects were developed, assessments were made about organizational structure, systems and tools, and training and education.
- Patrick Mills, Ken Evers, Donna Kinlin, and Robert S. Tripp, *Supporting Air and Space Expeditionary Forces: Expanded Operational Architecture for Combat Support Execution Planning and Control*, Santa Monica, Calif.: RAND Corporation, MG-316-AF, 2006. This monograph expands and provides more detail on several organizational nodes in our earlier work that outlined concepts for an operational architecture to guide the development of Air Force combat support execution planning and control needed to enable rapid deployment and employment of the Air and Space Expeditionary Force (AEF). These combat support planning, execution, and control processes are sometimes referred to as ACS C2 processes.
- Kristin F. Lynch, John G. Drew, Robert S. Tripp, and Charles Robert Roll, Jr., *Supporting Air and Space Expeditionary Forces: Lessons from Operation Iraqi Freedom*, Santa Monica, Calif.: RAND Corporation, MG-193-AF, 2005. This monograph describes expeditionary combat support experiences during the war in Iraq and compares these experiences with those associated with Joint Task Force Noble Anvil in Serbia and Operation Enduring Freedom (OEF) in Afghanistan. This monograph analyzes how combat support performed and how combat support concepts were implemented in Iraq, compares current experiences to identify similarities and unique practices, and indicates how well the combat support framework performed during these contingency operations.

- Don Snyder and Patrick Mills, *Supporting Air and Space Expeditionary Forces: A Methodology for Determining Air Force Deployment Requirements,* Santa Monica, Calif.: RAND Corporation, MG-176-AF, 2004. This monograph outlines a methodology for determining manpower and equipment deployment requirements. It describes a prototype policy analysis support tool based on this methodology, the Strategic Tool for the Analysis of Required Transportation (START), which generates a list of capability units called unit type codes that are required to support a user-specified operation. The program also determines movement characteristics. A fully implemented tool based on this prototype should prove to be useful to the Air Force in both deliberate and crisis action planning.
- James A. Leftwich, Robert S. Tripp, Amanda B. Geller, Patrick Mills, Tom LaTourrette, Charles Robert Roll, Jr., Cauley von Hoffman, and David Johansen, *Supporting Expeditionary Aerospace Forces: An Operational Architecture for Combat Support Execution Planning and Control*, Santa Monica, Calif.: RAND Corporation, MR-1536-AF, 2002. This report outlines the framework for evaluating options for combat support execution planning and control. The analysis describes the combat support C2 operational architecture as it is now and as it should be in the future. It also describes the changes that must take place to achieve that future state.
- Robert S. Tripp, Lionel A. Galway, Timothy L. Ramey, Mahyar A. Amouzegar, and Eric Peltz, *Supporting Expeditionary Aerospace Forces: A Concept for Evolving to the Agile Combat Support/Mobility System of the Future*, Santa Monica, Calif.: RAND Corporation, MR-1179-AF, 2000. This report describes a vision for the combat support system of the future based on individual commodity study results.
- Robert S. Tripp, Lionel A. Galway, Paul Killingsworth, Eric Peltz, Timothy L. Ramey, and John G. Drew, *Supporting Expeditionary Aerospace Forces: An Integrated Strategic Agile Combat Support Planning Framework*, Santa Monica, Calif.: RAND Corporation, MR-1056-AF, 1999. This report describes an integrated combat support planning framework that can be used to evaluate support options on a continuing basis, particularly as technology, force structure, and threats change.

RAND Project AIR FORCE

RAND Project AIR FORCE (PAF), a division of the RAND Corporation, is the U.S. Air Force's federally funded research and development center for studies and analyses. PAF provides the Air Force with independent analyses of policy alternatives affecting the development, employment, combat readiness, and support of current and future air, space, and cyber forces. Research is conducted in four programs: Force Modernization and Employment; Manpower, Personnel, and Training; Resource Management; and Strategy and Doctrine.

Additional information about PAF is available on our website:
http://www.rand.org/paf

Contents

Figures

Summary

There has always been disparity between the availability of combat support resources and process performance and the capabilities needed to support military operations. There are several reasons for this imbalance, including the inability to precisely predict resource requirements for contingency operations, inherent uncertainty in supply chain actions associated with providing combat support resources to the battlefield, unanticipated demands for resources to meet training and other operational requirements, and the development of budgets to meet estimated requirements several years in advance of when the monies become available. The current defense environment, characterized by budget pressures, the withdrawal from Iraq and Afghanistan, and a new defense strategy, will likely exacerbate the imbalance between the availability of combat support resources and requirements for them.

Because of these imbalances, operational commanders, the authorities who prioritize and allocate scarce resources among operational commanders, and resource providers need to know how combat support enterprise constraints and alternative resource allocation decisions would impact planned and potential operations. They also need to know when agile combat support (ACS) process performance breaches the control parameters set to meet contingency operation requirements.[1]

Previous RAND analyses found deficiencies in the Air Force ACS planning, execution, monitoring, and control processes that support Air Force operations.[2] The purpose of this analysis is to identify and describe where shortfalls or major gaps exist between current ACS processes (the AS-IS) and the vision for integrating enhanced ACS processes into Air Force command and control (C2) (the TO-BE) as presented in the operational architecture that we developed as part of this analysis.[3] We further suggest mitigation strategies needed to facilitate an efficient and effective global C2 network.

---

[1] By *control parameters* we mean a set level or acceptable threshold by which to track actual combat support performance so that, when a combat support parameter falls outside the set limits, combat support planners are notified so they can develop plans to bring the process back within control limits.

[2] Robert S. Tripp, Kristin F. Lynch, John G. Drew, and Robert DeFeo, *Improving Air Force Command and Control Through Enhanced Agile Combat Support Planning, Execution, Monitoring, and Control Processes*, Santa Monica, Calif.: RAND Corporation, MG-1070-AF, 2012.

[3] See Kristin F. Lynch, John G. Drew, Robert S. Tripp, Daniel M. Romano, Jin Woo Yi, and Amy L. Maletic, *An Operational Architecture for Improving Air Force Command and Control Through Enhanced Agile Combat Support Planning, Execution, Monitoring, and Control Processes*, Santa Monica, Calif.:

Research Approach

We began this analysis by evaluating RAND-developed operational architectures from 2002 and 2006.[4] We reviewed the recommendations of the previous analyses and evaluated Air Force progress in addressing the issues. We then evaluated how changes in the operational and fiscal environment affect ACS processes. The result is an updated operational architecture reflecting a vision for how enhanced ACS planning, execution, monitoring, and control processes could be integrated into Air Force and Joint C2 processes. Figure S.1 is a graphic depiction of the vision, highlighting the nodes that play a role in C2 processes.[5]

In the updated operational architecture, we outline the roles and responsibilities at each echelon—the President and Secretary of Defense (SECDEF), combatant commands (COCOMs), joint task forces (JTFs), component numbered Air Forces (C-NAFs), global ACS functional managers, supporting commands, units, and sources of supply—and across the phases of an operation—readiness, planning, deployment, employment and sustainment, and reconstitution.[6]

RAND Corporation, RR-261-AF, 2014. This report describes in detail a strategic- and operational-level C2 architecture integrating enhanced ACS processes.

[4] James Leftwich et al., *Supporting Expeditionary Aerospace Forces: An Operational Architecture for Combat Support Execution Planning and Control*, Santa Monica, Calif.: RAND Corporation, MR-1536-AF, 2002; and Patrick Mills et al., *Supporting Air and Space Expeditionary Forces: Expanded Operational Architecture for Combat Support Execution Planning and Control,* Santa Monica, Calif.: RAND Corporation, MG-316-AF, 2006.

[5] The DoD Architecture Framework (DoDAF), established as a guide for the development of architectures for the Department of Defense (DoD), defines a high-level graphic depiction of a concept as an Operational Viewpoint-1 (OV-1). Figure S.1 presents the OV-1 for the updated architecture.

[6] The details of the architecture are captured in both a visual representation developed using Microsoft Visio and in a spreadsheet developed using Microsoft Excel and can be found in Lynch et al. (2014).

Figure S.1
Vision for Enhanced ACS Processes

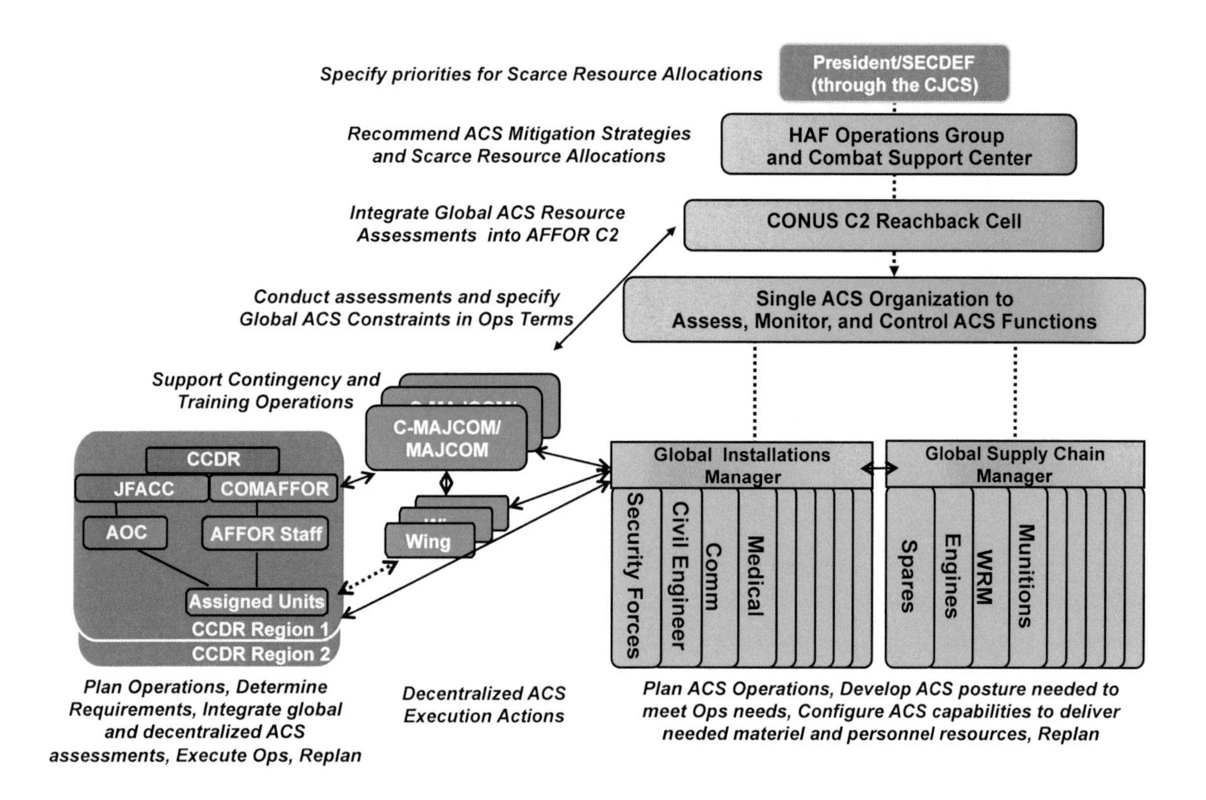

NOTES: Purple represents Joint processes, blue represents Air Force processes, green represents ACS processes, and orange represents processes defined in this architecture that are not currently assigned to a specific organization. Here we show ACS functions grouped by installation support and supply chain management. There are other ACS functions that fall outside these groupings (for example, chaplain, historian, acquisition, and test and evaluation) that also need to be managed globally and integrated with the other ACS functional capabilities to provide a complete picture of ACS capabilities and constraints.

For this analysis, we used the updated architecture to identify gaps and shortfalls in current processes. The architecture was designed to be broad enough to convey the vision but detailed enough to use to perform a doctrine, organization, training, materiel, leadership and education, personnel, and facilities (DOTMLPF) analysis to identify gaps and shortfalls that would prevent the Air Force from achieving the vision. The findings from our DOTMLPF analysis are documented here. After we identify current shortfalls, we present options for addressing them to help move the Air Force toward the vision presented in the updated TO-BE operational architecture.

Gaps Identified Using the Operational Architecture and Recommended Strategies to Enhance Command and Control

The concepts and processes we describe in the updated architecture have been widely vetted with senior operational and ACS leaders and there is agreement that the enhanced ACS processes are needed; however, the current ACS system does not fully support the vision shown in Figure S.1. There are gaps and shortfalls in many areas, including process; doctrine, guidance, and instructions; training and career management; and tools and systems.

Process

In the process area, the overarching shortfall is the inability to provide an enterprise assessment of combat support capabilities and constraints to inform trade-off decisions so that scarce resources can be effectively and efficiently used to meet Air Force operational priorities. This is a gap that spans across planning, execution, monitoring, and control—from the evaluation of deliberate plans through deployment, employment, and reconstitution, when plans are being executed, monitored, and controlled. To address this shortfall, processes need to be enhanced

- within individual ACS supply chains and functional capabilities: Global capabilities need to be assessed in a standard, repeatable manner that is linked directly to the ability to meet operational requirements.
- across individual supply chains and functional capabilities: The results of individual supply chain and functional capability assessments need to be integrated and balanced into a set of capabilities that can be used in planning (both deliberate and contingency) and replanning processes.
- within the Air Force: There should be a defined process to arbitrate between and among competing operational demands.

The Air Force has taken steps to address these process shortfalls. For example

- a few ACS functional capabilities (such as munitions) are managing capabilities and resources from an enterprise perspective, providing some visibility of worldwide capabilities and limitations
- C-NAFs were established as the Air Force component organizational structure to enhance operational-level C2 of air, space, and information operations across a broad range of engagements
- the logistics enterprise is being transformed so that it is both more responsive to combatant commander (CCDR) needs and more efficient in training, organizing, and equipping forces for operational missions

- the Air Force designated 12 service core functions to present warfighting capabilities to CCDRs and link resource requirements to needed operational capabilities in support of future programming requirements.[7]

However, there is much work still needed.

Doctrine, Guidance, and Instructions

Currently, doctrine and policy do not clearly define and delineate the C2 roles and responsibilities of combat support organizations. Standardized processes for identifying global resource capabilities, shortages, and operational outcomes associated with scarce resource allocation decisions should be established and defined in doctrine, guidance, and instructions to provide leaders with the information they need to make trade-off decisions.

While progress has been made in demonstrating how ACS planning, execution, monitoring, and control concepts enhance Air Force and Joint C2, there is still much work to be done to codify the processes in doctrine, guidance, and instructions, which would help to institutionalize process improvements.[8]

Training and Career Management

Trained personnel are necessary to help remedy shortfalls in the processes outlined above—conducting integrated capability assessments and developing scarce resource allocation schemes. It may be necessary to develop a new ACS planning, execution, monitoring, and control curricula, which could then be incorporated into existing or new training courses.

Training and realistic exercises are critical aspects of the link between combat support and operational planning. Educating both combat support and operations personnel about their roles in the context of campaign planning will enable more effective communication and facilitate the integrated decisionmaking process outlined in the operational architecture.

---

[7] For more information about recent Air Force process improvements as related to these identified shortfalls, see Chapters One and Two of Lynch et al. (2014).

[8] The Air Force C2 Integration Center has constructed experiments— the Agile Logistics EXperiment (ALEX) series—to test the concepts and learn more about how to implement the processes described in this architecture.

Recommendations

While there is general agreement about the need and value of the processes we describe, the responsibility for developing them spans many organizations. For instance, the C2 Core Function Team, which is responsible for developing C-NAF and component major command (C-MAJCOM) C2, views its programming support responsibility as the integration of information coming from the enhanced ACS processes (such as ACS constraint information) into the C-NAF and C-MAJCOM node for use in operational C2 (such as course of action [COA] selection). However, it is the responsibility of the ACS Core Function Team to program for developing and sustaining decision support systems to produce that information.

One way for the Air Force to begin addressing these challenges is to call a C2 symposium to bring together Air Force communities that play key roles in this area. The symposium could be used to define assessment and control technique requirements; vet capabilities; identify necessary changes in doctrine, guidance, and instructions; develop needed training enhancements; and define organizational roles and responsibilities.

The architecture developed as a companion to this analysis may be of particular importance if enhanced ACS processes are developed separately by several organizations (such as the Air Force Materiel Command [AFMC], the Air Education and Training Command [AETC], and the Air Combat Command [ACC]) and the information from these processes is integrated at C2 nodes at the operational and strategic levels by other organizations (such as the Air Staff). The architecture identifies the processes that can be developed by one organization and the outputs that can be integrated into the processes developed by other organizations. In short, it does not matter who develops the processes and associated systems from an architecture perspective.

While progress has been made in improving ACS planning, execution, monitoring, and control processes, there are additional actions that need to be taken. Because the issues are broader than any one Air Force organization, closing the gaps may be difficult without designating a single leader as the ACS planning, execution, monitoring, and control commander. A single ACS organization should be given responsibility for and the authority to address the gaps outlined above.

Establishing a single ACS authority would be a large cultural and organizational shift for the Air Force that would take time to implement. In the meantime, there are many actions that can be taken to improve ACS planning, execution, monitoring, and control. Specifically, to institutionalize and further enhance the integrated assessment and allocation processes discussed in this report, the Air Force should continue to experiment and participate in exercises that can demonstrate and enhance these processes over time. For example, we recommend that the Air Force continue the ALEX series of experiments using an ACS reachback cell at an operational support facility (OSF) to conduct global assessments and provide capability and constraint feedback to C-NAF and C-MAJCOM planners. The ACS reachback cell could also provide analysis to planners during exercises, such as Austere Challenge, Terminal Fury, and Ulchi Freedom Guardian. Key to these assessments is an organization with tools and trained personnel that is tasked with supporting C-NAF and C-MAJCOM planning, exercises, and experiments.

The enhanced ACS processes the Air Force implements need to be codified in doctrine, guidance, and instructions. The roles and responsibilities of each C2 node, including logistics, operational, and installation staff; Air Force commanders; major commands (MAJCOMs), specifically AFMC; and others, such as an ACS reachback cell, should be delineated. Specifically, the logistics sustainability analysis (LSA) process defined in AFI 10-401[9] should be updated to require global integrated ACS resource assessments and prioritization rules for allocating scarce resources.

Once Air Force–level guidance defines and assigns roles and responsibilities, AFMC and the Air Force Sustainment Center (including the Air Logistics Complexes [ALCs]) need to develop corresponding instructions outlining command organizational roles in C2 processes, such as

- the command role in proactive operation plan (OPLAN) assessments and contingency planning (modify Air Force Materiel Command Instruction [AFMCI] 10-204)[10]
- the roles of the Air Force Sustainment Center (which includes the former Air Force Global Logistics Support Center [AFGLSC]) and Headquarters in developing proactive risk mitigation strategies (modify the Headquarters and ALC OPLAN 70s)
- a single point of contact (POC) to direct guidance for AFMC across OPLANs, contingency operations, exercises, experiments, and wargame C2 responsibilities.

The Air Force has moved forward in achieving the vision presented in the operational architecture, but our research shows that many actions still can be taken to improve ACS planning, execution, monitoring, and control processes; doctrine, guidance, and instructions; training and career management; and tools and systems.

---

[9] U.S. Air Force, Air Force Instruction 10-401, *Air Force Operations Planning and Execution*, December 7, 2006.

[10] U.S. Air Force, Air Force Materiel Command Instruction 10-204, *AFMC Exercise Program*, August 31, 2010b.

Acknowledgments

Numerous people both within and outside of the Air Force provided valuable assistance to and support of our work. They are listed here with their rank and position as of the time of this research. We thank Lt Gen Terry Gabreski, AFMC Vice Commander (AFMC/CV); Lt Gen Janet Wolfenbarger, AFMC/CV; and Lt Gen Loren Reno, Deputy Chief of Staff for Logistics, Installations, and Mission Support (AF/A4/7), for sponsoring this work. We also thank their staffs for their time and support during this research.

This work would not have been possible without the support of many individuals. At the Air Staff, we thank Patricia Young, AF/A4/7; Grover Dunn, Director of System Integration, Deputy Chief of Staff, Logistics, Installations, and Mission Support (AF/A4I); David Beecroft, Col Patricia Battles, Dick Olson, Freddie McSears, Laine Krat, and Robert Ekstroem from the Directorate of Global Combat Support, Deputy Chief of Staff, Logistics, Installations, and Mission Support (AF/A4/7Z); Maj Gen Richard Devereaux, Deputy Chief of Staff, Operations, Plans and Requirements, Director of Operational Planning, Policy, and Strategy (AF/A5X); and Allen Wickman, Directorate of Operations for the Deputy Chief of Staff, Operations, Plans, and Requirements (AF/A3O). Also at the Air Staff, we thank Chief Scott Heisterkamp, John Ray, Michael Robertson, William (Dave) Sweet, Nick Reybrock, Col Jeffery Vinger, Col James Iken, Roy Bousquet, Lt Col Paul Story, Kevin Allen, Col Tracy Tenney, Lt Col Edward Lagrou, and Keith Tucker for their time discussing and reviewing our architecture analysis.

At AFMC, we thank Maj Gen David Eidsaune, Air Force Materiel Command, Director, Strategic Plans, Programs and Analyses (AFMC/A8/9); Richard Moore; Bob McCormick; Tom Stafford; Col Chris Froehlich; Lt Col Kendra Eagan; William Santiago; Molly Waters; Col Arley Hugghins; and Lt Col Carl Myers. From the AFGLSC, we thank Maj Gen Gary McCoy, Commander, and Brig Gen Brent Baker, Commander, as well as Lorna Estep, Col Mark Johnson, Col Ray Lindsay, Richard Reed, Michael Howenstine, Col Jeffrey Sick, Lt Col Kevin Gaudette, William (Steve) Long, Mike Niklas, Frank Washburn, Debra Garves, Mel Cooper, James Weeks, and Lynne Grile.

In the Pacific area of responsibility (AOR), we thank Lt Gen Hawk Carlisle, Commander, 13th Air Force (13AF); Col Gregory Cain, 13AF Chief of Staff; and Col Darlene Sanders, 13AF Director of Logistics (13AF/A4); as well as Elaine Ayers, Capt James Arnett, and the entire 13AF staff for their time and cooperation. At Pacific Air Forces (PACAF), we thank Maj Gen Jan-Marc Jouas and Brig Gen Michael Keltz,

PACAF A3/5/8; Col Joseph Martin, PACAF/A4; Col Karl Bosworth, PACAF/A7; Donald Casing; Russell Grunch; and Capt Jolie Gibbs. And, at 7th Air Force, we thank Lt Gen Jeffrey Remington, Commander, and his entire staff for their time.

In the European AOR, we thank Lt Gen Frank Gorenc, Commander, 3rd Air Force (3AF); Col Raymond Strasburger, 3AF/CoS; Col Darrell Mosley and Lt Col Manuel (Paul) Perez, 3AF/A4; Mr. Phillip Romanowicz, 3AF/A9; and the entire 3rd Air Force staff. At 17th Air Force, we thank Maj Gen Margaret Woodward, Commander; Brig Gen Michael Callan, 17AF/CV; Col Chris Hair, 17AF/CoS; Maj Jason Barnes; and the rest of the staff. At U.S. Air Forces, Europe, we thank Brig Gen John Cooper, USAFE/A4/7, and Eric Jacobson.

At 12th Air Force, we thank Brig Gen Jon Norman, 12AF/CV; Col Kyle Ingham; Col Byron Mathewson; Thomas Schnee; Lt Col John Landolt; and the entire staff. At Air Forces Central (AFCENT), we thank Brig Gen Richard Shook, AFCENT/CCA, and the staff. At ACC, we thank Maj Gen Judith Fedder, ACC Director of Logistics; Curtis Gibson; Robert Potter; Col David Crow; MSgt Richard Amann; and MSgt Jeremy Yates. At the Air Force Command and Control Integration Center, we thank Stan Newberry, Director; Col David Baylor; Col Brian Pierson; Lt Col Darrell Pennington; Capt Marc Morin; Capt Jaylene Pombrio; Desiree Stone; and Thomas Connors. At Air Mobility Command (AMC), we thank Kevin Beebe, Maj James Donelson, and Craig Harris. On the Secretary of Air Force (SECAF) staff, we thank Mike McFarren, Col David Geuting, and Deborah Dewitt. At the Joint Staff, we thank Susan McDonald. At the Vehicle and Equipment Management Support Office, we thank SMSgt Scotty Browning. At the Air Force Civil Engineer Support Agency (AFCESA), we thank Gregory Cummings, Dennis Cook, and the entire staff. Finally, at the Ogden Air Logistics Center, we thank Mark Johnson, Col Perry Oaks, Mark Brown, Wendy Kierpiec, and the entire Global Ammunition Control Point staff for their time.

At RAND, we are grateful for the support given by John Ausink, Laura Baldwin, Ed Chan, and Lt Col Peter Breed. We would especially like to thank Ron McGarvey and Don Snyder for their thorough review of this report. Their reviews helped shape it into its final, improved form.

Responsibility for the content of the document, analyses, and conclusions lies solely with the authors.

Abbreviations

638 SCMG	638th Supply Chain Management Group
710 COS	710 Combat Operations Squadron
A3/5	Air, Space, and Information Operations Directorate
A4	Logistics Directorate
A7	Installations and Mission Support Directorate
ACC	Air Combat Command
ACS	agile combat support
ACS C2	agile combat support command and control
AEF	Air and Space Expeditionary Force
AETC	Air Education and Training Command
AF/A3/5	Deputy Chief of Staff, Operations, Plans and Requirements
AF/A4/7	Deputy Chief of Staff, Logistics, Installations, and Mission Support
AFB	Air Force Base
AFC2IC	Air Force Command and Control Integration Center
AFCENT	Air Forces Central
AFEUR	Air Forces Europe
AFFOR	Air Force forces
AFGLSC	Air Force Global Logistics Support Center
AFI	Air Force Instruction
AFIT	Air Force Institute of Technology
AFMC	Air Force Materiel Command
AFMC/A3	AFMC Directorate of Air, Space, and Information Operations
AFMC/A4	AFMC Logistics Directorate
AFMC/A3X	AFMC Operational Plans Division
AFMC/A8XI	AFMC Wargaming Integration Office
AFMC/A8XW	AFMC Wargaming Integration Office
AFMCI	Air Force Materiel Command Instruction
AFMCSUP I	AFMC Supplement 1
AFPD	Air Force Policy Directive
AFRC	Air Force Reserve Command
AFSC	Air Force specialty code
ALC	Air Logistics Complex
ALEX	Agile Logistics EXperiment

AMC	Air Mobility Command
ANG	Air National Guard
AOC	Air and Space Operations Center
AOR	area of responsibility
ART	AEF Reporting Tool
BEAR	basic expeditionary airfield resources
BES	Budget Estimate Submission
BSP	base support plan
C2	command and control
CAT	Crisis Action Team
CC	Commander
CCDR	combatant commander
CE	civil engineering
CJCSI	Chairman of the Joint Chiefs of Staff Instruction
CJCSM	Chairman of the Joint Chiefs of Staff Memorandum
C-MAJCOM	component major command
C-NAF	component numbered Air Force
C-NAF/CC	component numbered Air Force commander
COA	course of action
COCOM	combatant command
COMAFFOR	Commander of Air Force Forces
CONOPS	concept of operations
CONPLAN	contingency plan
COOP	continuity of operations
COS	Combat Operations Squadron
CRP	Contract Repair Process
CSAF	Chief of Staff, United States Air Force
C-SPEC	combat support planning, execution, and control
CSSC	COMAFFOR Senior Staff Course
DEPORD	deployment order
DLA	Defense Logistics Agency
DoD	Department of Defense
DoDAF	DoD Architecture Framework
DOTMLPF	doctrine, organization, training, materiel, leadership and education, personnel, and facilities
DPG	Defense Planning Guidance
DREP	Depot Repair Enhancement Program
DRU	direct reporting unit

ECS	expeditionary combat support
eLog21	Expeditionary Logistics for the 21st Century
ESP	expeditionary site plan
ESSP	Expeditionary Site Survey Process
EUCOM	U.S. European Command
EXPRESS	Execution and Prioritization of Repair Support System
FAM	Forward Operating Location Assessment Model
FOC	full operational capability
FOL	forward operating location
GACP	Global Ammunition Control Point
GCCS	Global Command and Control System
GCSS	Global Combat Support System
GFM	global force management
GIC	global integration center
GSU	Geographically Separated Unit
ICC	Installation Control Center
IDEF	Integrated Definition
IG	inspector general
IGESP	In-Garrison Expeditionary Site Plan
IM	Item Management
IOC	initial operational capability
JCS	Joint Chiefs of Staff
JEFX	Joint Expeditionary Force Experiment
JOPES	Joint Operation Planning and Execution System
JSCP	Joint Strategic Capabilities Plan
JSPS	Joint Strategic Planning System
JTF	joint task force
LFF	Logistics Factors File
LOGCAT	Logistician's Contingency Assessment Tools
LSA	logistics sustainability analysis
MAJCOM	major command
MCC	Materiel Control Center
NSCS	National Security Council System
OC-ALC	Oklahoma City Air Logistics Complex
OCR	office of coordinating responsibility
OO-ALC	Ogden Air Logistics Complex
OPLAN	operation plan
OPR	office of primary responsibility

OPSO/A3XC	AFMC Command Center
OPT	operational planning team
OSC	operational support center
OSD	Office of the Secretary of Defense
OSF	operational support facility
OV	Operational Viewpoint
PACAF	Pacific Air Forces
PACOM	U.S. Pacific Command
PAD	Program Action Directive
PAF	Project AIR FORCE
POC	point of contact
POM	Program Objective Memorandum
PPBE	planning, programming, budgeting, and execution
RAT	Rapid Augmentation Team
SAF/US(M)	Secretary of the Air Force, Business Transformation and Management Office
SCMG	supply chain management group
SECDEF	Secretary of Defense
SF	security forces
SOR	Source of Repair
SPD	System Program Director
START	Strategic Tool for the Analysis of Required Transportation
TPFDD	time-phased force and deployment data
UTC	Unit Type Code
WAAR	wartime aircraft activity report
WMP-1	War and Mobilization Plan Volume 1
WMS	Wartime Materiel Support
WR-ALC	Warner Robbins Air Logistics Complex
WRM	war reserve materiel

1. Introduction, Background, and Motivation

Air Force Doctrine Document 1 states that command and control (C2) of air, space, and cyber power is a fundamental function of the United States Air Force.[1] C2 enables the United States military to conduct operations that accomplish specific military objectives. Agile combat support (ACS),[2] another fundamental function of the Air Force, plays an integral role in C2. Often referred to as *agile combat support command and control* (ACS C2), the planning, execution, monitoring, and control of ACS processes are an integral part of Air Force and Joint C2. Prior Project AIR FORCE (PAF) research[3] found that ACS planning, execution, monitoring, and control processes critical to informing C2 decisions are not adequately defined and delineated in doctrine, guidance, and instructions, and tools or systems to support these ACS processes are lacking. The purpose of this analysis is to identify and describe where shortfalls or major gaps exist between current ACS processes (the AS-IS) and the vision for integrating enhanced ACS processes into Air Force C2 (the TO-BE) as presented in the operational architecture that we developed as part of this analysis and is presented in the companion report.[4] We further suggest mitigation strategies to facilitate an efficient and effective global C2 network.

---

[1] U.S. Air Force, Air Force Doctrine Document 1, *Air Force Basic Doctrine*, November 17, 2003.

[2] In this document, the term *ACS* refers to the 26 functional capabilities outlined in U.S. Air Force, *Agile Combat Support Command and Control (ACS C2) Supporting CONOPS*, November 15, 2008, p. 10, Figure 2. ACS is broader than just logistics; it includes personnel, services, communications, and installation and mission support functions, just to name a few.

[3] James Leftwich et al., *Supporting Expeditionary Aerospace Forces: An Operational Architecture for Combat Support Execution Planning and Control*, Santa Monica, Calif.: RAND Corporation, MR-1536-AF, 2002; Patrick Mills et al., *Supporting Air and Space Expeditionary Forces: Expanded Operational Architecture for Combat Support Execution Planning and Control,* Santa Monica, Calif.: RAND Corporation, MG-316-AF, 2006; and Robert S. Tripp et al., *Improving Air Force Command and Control Through Enhanced Agile Combat Support Planning, Execution, Monitoring, and Control Processes*, Santa Monica, Calif.: RAND Corporation, MG-1070-AF, 2012.

[4] See Kristin F. Lynch, John G. Drew, Robert S. Tripp, Daniel M. Romano, Jin Woo Yi, and Amy L. Maletic, *An Operational Architecture for Improving Air Force Command and Control Through Enhanced Agile Combat Support Planning, Execution, Monitoring, and Control Processes*, Santa Monica, Calif.: RAND Corporation, RR-261-AF, 2014. This report describes, in detail, a strategic- and operational-level C2 architecture integrating enhanced ACS processes.

Background and Research Motivation

There has always been disparity between the availability of combat support resources and process performance and the capabilities needed to support military operations. There are many factors that contribute to this imbalance between needed ACS resources and those available for contingency and training operations. These factors include the inability to precisely predict resource requirements, the development of budgets to meet estimated requirements several years in advance of when the monies become available, the inherent uncertainty in supply chain actions associated with providing combat support resources to the battlefield, the potential need to reallocate funding to meet unanticipated requirements, and unforeseen world events that present new and emerging requirements. The current defense environment, characterized by budget pressures, the withdrawal from Iraq and Afghanistan, and new defense strategy, will likely exacerbate the imbalance between the availability of combat support resources and requirements for them.

Simultaneously, there is increasing pressure to conduct all Department of Defense (DoD) operations more efficiently to accommodate reduced budgets over time. Within the context of C2, this means providing quick, tailorable support packages optimized to meet specific operational needs. Economic pressures are likely to continue and may result in further reductions in the resources available to support military operations.

Component numbered Air Force commanders (C-NAF/CCs), component major command commanders (C-MAJCOM/CCs), and their staffs develop contingency courses of action (COAs) with limited information about global ACS resource availabilities and constraints. The assumption that sufficient ACS resources exist to simultaneously meet all worldwide operational priorities is not credible. As a result, C-NAF/CCs, C-MAJCOM/CCs, and their staffs do not fully understand or anticipate the risks associated with specific COAs, and they do not take steps (such as changes in operational plans or ACS plans) in advance to mitigate those risks.

To meet future operational requirements with the limited resources available, the President and the Secretary of Defense (SECDEF) may need to allocate scarce resources among competing demands. Individual ACS supply chain managers and functional resource managers need to be integrated into important C2 nodes to provide enterprise-wide assessments of ACS capabilities and constraints. The ACS community needs to be able to provide predictions of combat support needs and rapid responses to dynamic operational needs, as well as allocate scarce resources to where they are most needed as determined by the President's/SECDEF's priorities.

Today, most ACS planning, execution, monitoring, and control processes are ad hoc. Only a few functional capabilities manage resources from an enterprise perspective. Rather, many ACS resources are viewed from a more narrow theater perspective without an enterprise view of worldwide capability. Still, combat support of military operations remains successful primarily because of the efforts of individuals in the combat support community to overcome the difficulties of current (AS-IS) processes, systems, tools, organizations, and training. Since the Air Force will continue to operate in a resource-constrained environment in the future, standard, repeatable analytic ACS processes to support trade-off and allocation decisionmaking should be established and implemented.

The Air Force and DoD recognize the need to change to meet existing and emerging global requirements with limited resources. Both have made significant investments in improving the capabilities needed to meet the challenges posed by the current defense environment.[5] The Air Force has begun to transform its logistics enterprise so that it is both more responsive to combatant commander (CCDR) needs and more efficient at training, organizing, and equipping forces for operational missions.[6]

In light of these recent Air Force transformations and changes in the operational environment, in 2009 senior Air Force logisticians asked PAF to examine ACS processes to meet contingency, readiness preparation, and training requirements. Specifically, we were asked to review prior RAND-developed operational architectures, identify necessary changes resulting from the transformational efforts, and evaluate whether the gaps and shortfalls identified in previous work were still present. The previous analysis found that the Air Force lacked the comprehensive doctrine, guidance, and instructions; processes; organizations; training; and tools and systems needed to enable combat support functions to allocate and utilize limited resources to best achieve operational objectives both in contingency operations and during readiness preparation and training.[7] To address some of those issues, we recommended that standardized assessments of global ACS capabilities and constraints be included in contingency planning and execution activities.

---

[5] For example, Program Action Directive (PAD) 06-09 established component numbered Air Forces (C-NAFs) as the Air Force component organizational structure to enhance operational-level C2 of air, space, and information operations across a broad range of engagements (U.S. Air Force, Program Action Directive 06-09, *Implementation of the Chief of Staff of the Air Force Direction to Establish an Air Force Component Organization*, November 7, 2006b; and U.S. Air Force, *Air Force Forces Command and Control Enabling Concept*, Change 2, May 25, 2006a); the Air and Space Operations Center (AOC), a part of the C-NAF, was designated as a weapon system whose process-oriented focus is on producing war plans and executing them to achieve strategic and tactical objectives.

[6] For example, the Expeditionary Logistics for the 21st Century (eLog21) program of initiatives was developed to modernize and streamline logistics operations to address the challenges of this more demanding environment within limited budgets. ELog21 is an umbrella program comprising many different logistics and supply chain transformational initiatives with the overall goal of improving resource availability and reducing costs.

[7] Robert S. Tripp et al., 2012.

In the companion report, we expand the previous RAND work that identified shortfalls limiting the Air Force's ability to allocate and utilize limited combat support resources to best achieve operational objectives and present an updated operational architecture that outlines a vision for integrating enhanced ACS processes into Air Force C2 at the strategic and operational levels in light of the current defense environment.[8] We document the ACS processes needed to work within the Air Force and Joint C2 enterprise to help the warfighter achieve the desired operational effects. Here, we use that operational architecture to identify gaps and shortfalls in current ACS processes and suggest options for how to close those gaps.

Organization of This Report

In the chapters that follow, we identify shortfalls and gaps in current ACS processes using the updated operational architecture. In Chapter Two we describe the gaps and shortfalls that exist today in processes; doctrine, guidance, and instructions; training and career management; tools and systems; and organizations. We also suggest strategies for closing those gaps. Chapter Three concludes with recommendations for improved ACS planning, execution, monitoring, and control.

In addition, there are two appendixes to this document. Appendix A discusses the Agile Logistics Evaluation EXperiment and Appendix B lists relevant Air Force and DoD doctrine, guidance, and instructions with recommended changes and additions needed to address the ACS vision presented in this document.

[8] See Lynch et al. (2014) for the detailed architecture developed as part of this analysis.

2. Gaps and Shortfalls Identified Using the Operational Architecture and Recommended Strategies to Enhance Command and Control

We began this analysis by evaluating previous RAND-developed operational architectures from 2002 and 2006.[1] Then, we refined the previous work in light of the current operational and fiscal environments and developed an updated architecture. The updated architecture outlines roles and responsibilities at each echelon—the President and SECDEF, combatant commands (COCOMs), joint task forces (JTFs), C-NAFs, global ACS functional managers, supporting commands, units, and sources of supply—and across the phases of an operation—readiness, planning, deployment, employment and sustainment, and reconstitution.[2] The architecture presents a vision for integrating enhanced ACS processes into Air Force C2 at the strategic and operational levels, with a single ACS integrator bringing together and balancing individual stovepiped ACS processes to provide capability and constraint assessments to senior leaders for priority and allocation decisionmaking, as shown in Figure 2.1.[3]

---

[1] Leftwich et al., 2002; Mills et al., 2006.

[2] Details on the development of the updated architecture and on the architecture itself can be found in Lynch et al. (2014).

[3] The TO-BE vision presented in Figure 2.1 shows a single organization responsible for integrating and balancing ACS functions. Other organizational options are outlined later is this section.

Figure 2.1
Vision for Enhanced ACS Processes

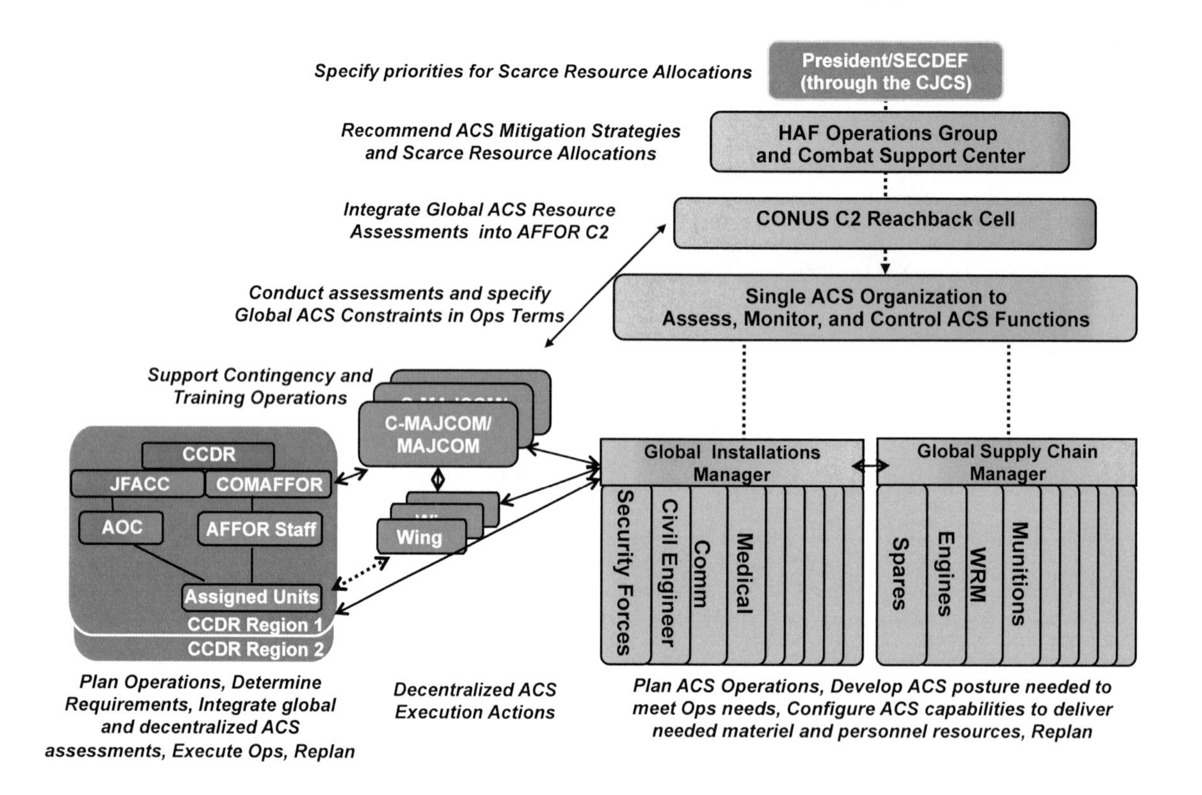

NOTE: Purple represents Joint processes, blue represents Air Force processes, green represents ACS processes, and orange represents processes defined in this architecture that are not currently assigned to a specific organization.

In Figure 2.1, we show only some of the ACS functions grouped by installation support and supply chain management. There are other ACS functions (for example, chaplain, historian, acquisition, and test and evaluation) that fall outside these groupings but also need to be managed globally and integrated with other ACS functional capabilities to provide a complete picture of ACS capabilities and constraints.

This vision for how the Air Force C2 system could work in the future has been well vetted with senior Air Force leadership; however, the current ACS system does not fully support the vision. Using the updated architecture developed as part of this analysis, we performed a doctrine, organization, training, materiel, leadership and education, personnel, and facilities (DOTMLPF) analysis to uncover any gaps or shortfalls that would prohibit the vision from being achieved. We found gaps and shortfalls in many areas. This chapter discusses the main process gap, as well as associated shortfalls in processes, doctrine, guidance and instructions, training and career development, organization, tools, and systems. We further recommend mitigation strategies to help close these gaps.

Agile Combat Support Command and Control Processes

The overarching process gap is the inability to provide an enterprise assessment of combat support capabilities and constraints that can be used to inform trade-off decisions so that scarce resources can be effectively and efficiently allocated to meet Air Force operational priorities. This gap spans across planning, execution, monitoring, and control—from evaluation of deliberate plans, during which a Logistics Sustainability Analysis (LSA) should inform the COCOM of capabilities and constraints, to a contingency for which COAs are developed and forces are deployed, employed, and reconstituted. There are shortfalls on several different levels associated with this gap:

- within individual ACS supply chains and functional capabilities: There is no standard, repeatable process to plan, execute, monitor, and control ACS supply chains and functional capabilities within the Air Force C2 system to proactively manage scarce ACS resources across competing operational demands.
- across individual supply chains and functional capabilities: Methods of combining individual supply chain and functional capability assessment results in an integrated and balanced set of capabilities that can be used in planning (both deliberate and contingency) and replanning processes are incomplete and there is no organization tasked with this responsibility.
- within the Air Force: Processes to arbitrate between and among competing operational demands are deficient.

We will address each of these shortfalls in the sections below.

Resource Assessment Processes Within Individual ACS Supply Chains and Functional Capabilities

Because of the funding limitations being imposed on the Program Objective Memorandum (POM) process, resource constraints are inevitable. There are not, nor will there be in the future, enough resources for CCDRs to have everything they need to support all operations in their areas of responsibility (AORs) as envisioned in Office of the Secretary of Defense (OSD) planning guidance. Resources must be shared globally to meet all demands. Global management of resources facilitates the allocation and reallocation of scarce resources as worldwide priorities shift.

Global supply chain and functional capability managers are needed to provide independent assessments of worldwide capabilities and constraints for each ACS resource.[4] In deliberate planning, this assessment could inform CCDRs during the LSA process. These assessments could also help inform the COA development process during contingency planning. Global managers should also be able to effectively and efficiently shift resources to where they are needed most.

Since 2002, when the first architecture was developed, the Air Force has taken steps to improve processes within individual supply chains by designating global managers for some resources: munitions at the GACP, select spare parts at the Air Force Global Logistics Support Center (AFGLSC) (which is now part of the Air Force Sustainment Center), and non-unit war reserve materiel (WRM) at Air Combat Command (ACC). Munitions, for example, has a global requirements determination process and an allocation board for distributing assets worldwide. In the fuels area, the Air Force has business rules and tools and systems that allow for worldwide planning. Logistics personnel input the types and number of aircraft, their expected usage during the contingency, and expected beddown locations into the fuels planning tool, Integrated Consumable Item Support, and it calculates the fuels requirement by location.[5] It is a well-defined and easy-to-use system employed throughout DoD to provide consistent requirements estimates. In addition, the Air Force has functional capability managers that are responsible for developing ACS personnel's skills and career path advancement—global managers for ACS personnel.[6]

In these areas, resources are being managed from an enterprise perspective. However, ACS planning, execution, monitoring, and control processes are ad hoc. Other resources, such as services and vehicles, do not have such well-defined, standardized, repeatable processes. They are managed theater by theater, without a global manager to integrate information into an enterprise view of worldwide capability.

The Air Force will continue to operate in a resource-constrained environment in the future. Therefore, leaders will need to make tough trade-off decisions when allocating scarce resources. The ACS system does not currently support trade-off and allocation decisionmaking with standardized, analytic processes for identifying global resource capabilities and shortages and the operational outcomes associated with scarce resource allocations. Global supply chain and functional capability managers should be established

---

[4] Then, those individual stovepiped assessments can be combined to provide an integrated assessment of the ability to meet Air Force operational priorities.

[5] Integrated Consumable Item Support is a Defense Logistics Agency (DLA) decision support system that can calculate the deployment requirements for over two million DLA-stocked items using time-phased force and deployment data.

[6] The global managers for personnel primarily manage home-station requirements rather than expeditionary requirements, but having a global manager is a step in the right direction.

for each resource area. Processes for assessing enterprise capability and shortfalls (such as requirements determination processes and real-time asset management) should be defined, practiced, and codified in doctrine, guidance, and instructions. Tools and systems should be developed to aid in these processes. Resources that are centrally managed and shared globally may be better suited to meet uncertain future demands.

Integrating and Balancing Across Individual Supply Chains and Functional Capabilities

As previously stated, individual resources are currently managed and controlled independently—some theater by theater and some globally. However, there is little integration across supply chains or among functional capabilities. Currently, there is no organization tasked with the responsibility to bring the individual stovepiped resources together into an integrated view of ACS capability.[7] To provide senior leaders with better visibility into global combat support capabilities and constraints, improvements are needed in combat support analytic assessment capabilities, metrics, and the organizational construct used to support these processes. We discuss each of these in the sections below.

Integrated Analytic Assessment Capabilities

Currently, independent, stovepiped resource assessments are not combined with other resource assessments to provide an integrated and balanced view of Air Force capability. For example, assessments of individual materiel resources, including WRM; vehicles and special purpose support equipment; munitions; petroleum, oil, and lubricants (POL); spare parts; personal equipment; and others are not integrated to determine how the resources interrelate in terms of affecting operational objectives. These materiel resources could be combined with other combat support resources, such as civil engineering, communications, and security forces (SF) capabilities, to provide an integrated assessment of meaningful data to operational planners.

Some functional capabilities are already conducting stovepiped assessments. For example, the Air Force Sustainment Center (which includes the former AFGLSC) already provides global assessments of spare parts for the Air Force. Their charter could be expanded to include all classes of materiel, making them a global manager for materiel. The Air Force Personnel Center (Directorate of Air and Space Expeditionary Force Operations) or another organization could be the global manager for personnel. For

---

[7] ALEX demonstrated this type of integrated assessment for ACS capabilities for a set of individual resources. ALEX is discussed in this chapter and in Appendix A.

integrated assessments to provide useful information to senior leaders, they should focus on operationally relevant metrics and show how ACS could support or constrain Air Force capabilities.

Metrics

To determine the combat support system's performance in terms of operationally relevant metrics, it is necessary to understand how materiel and non-materiel resources interact to produce desired capabilities. This is not currently done. Because these capability metrics depend upon more than just materiel, materiel managers need to do more than simply monitor the numbers of physical assets available in each category; they also need to understand how asset location, condition, and quantities interact with repair, if applicable, and how transportation times in each category contribute to operational effectiveness. Ideally, those responsible for understanding combat support resources would be able to relate the different categories of resources—materiel, infrastructure, personnel, and transportation[8]—to each other to determine the marginal contribution of individual resources against system-wide operational effectiveness output measures. Decisionmakers would then be positioned to make the most cost-effective use of combat support resources, maximizing the capability of a given set of resources to support the warfighter.

Metrics should also be based on the priorities laid out in the CCDR guidance. If the CCDR's goal is bombs on target, then the ACS metric may relate ACS resources to the ability to generate sorties or the ability to open bases from which to conduct operations.

The ACS community has the ability to relate some combat support resource levels and some processes' performance to operationally relevant metrics, such as mission generation capability, forward operating location (FOL) initial operational capability (IOC), and full operational capability (FOC), metrics used when developing COAs. For example, the sortie generation capability is a function of many combat support parameters, including the removal rates of avionics components; the maintenance throughput of the repair facility (both on base and at a repair facility); and movement capacity and throughput capability (for example, airlift frequency between the repair facility and a deployed location and the transportation time for these components). Degradation in any one of those combat support parameters will affect sortie generation capabilities, and the sorties projected may not meet the requirement.

---

[8] AFI 10-401 identifies requirements for conducting logistics supportability analyses to include assessments of materiel, infrastructure (usually focused on FOL ramp, runway, and other construction needs), combat support forces (usually focused on personnel issues associated with filling combat support unit type codes [UTCs]), and lift.

Ultimately, the goal should be to determine how alternative resource allocations impact bombs on target or other desired effects. In the meantime, several operationally relevant metrics, such as the ability to generate desired missions, the ability to establish and sustain a desired number of FOLs, the ability to provide required security, the ability to evacuate specific numbers of wounded or sick, and so forth, can help guide the allocation of scarce resources. The analysis of these metrics provides meaningful data to operations planners for any necessary replanning caused by constraints in ACS capabilities.

To conduct integrated and balanced capability assessments, models and tools are needed to help relate combat support resource levels and process performances to operationally relevant metrics. And, trained and assigned personnel who know how to use available models to access the relevant and authoritative data and to identify constraints in global resource availabilities are needed to perform the integrated assessments. A C2 symposium that brings together Air Force communities that play key roles in this area could be used to define and vet capabilities and develop needed training enhancements.

Centralized Management Authority

As pointed out previously, the Air Force has acted to address some of these capability assessment shortfalls by creating some global ACS organizations (for example, the Global Ammunition Control Point [GACP] for munitions and the AFGLSC [now part of the Air Force Sustainment Center] for spares). In Figure 2.2, we show the independent, stovepiped supply chains and functional capabilities in green, some of which have global managers (shown along the bottom of Figure 2.2). Efforts to establish global resource managers are a step in the right direction, as the global managers have improved the visibility of ACS manpower, equipment, and other materiel; however, they have stovepiped resource responsibilities and there is no defined process for combining or integrating individual resource assessments.

Figure 2.2
AS-IS—No Organization Provides Integrated ACS Capability Assessments

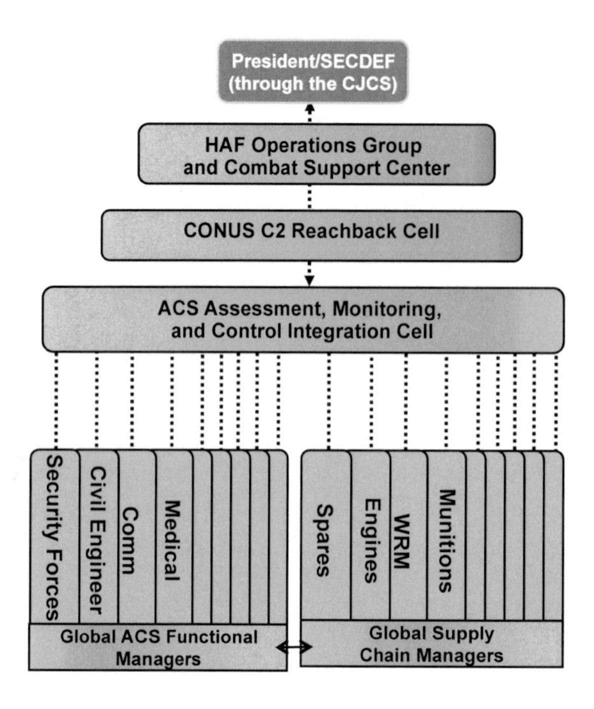

NOTE: Processes we defined in this analysis that are not currently assigned to an organization are shown in orange.

The Air Force lacks an enterprise organization with the analytic capability to identify integrated global ACS resource capabilities and constraints (for example, integrating munitions with spares and maintenance Unit Type Codes [UTCs] to assess sortie generation capability at a given location), including the ability to identify the most binding constraints with respect to specific COAs. Very few, if any, integrated assessments are conducted to support Air Force planning. Commanders of the Air Force forces (COMAFFORs), as a result, might be presenting COAs to Joint services and COCOMs that are not supportable from an Air Force global resource point of view.

The vision presented in this analysis calls for a single ACS organization to be responsible for bringing together these assessments from independent stovepipes and/or from global managers (shown as the "Single ACS Organization" in Figure 2.3). Each individual resource needs to be integrated and balanced to give an overall picture of ACS capability. Currently, such integration does not occur and no organization is tasked with the responsibility or has the authority to manage and control ACS resources across stovepipes. This type of integrated analysis is critical to the management and control of resources necessary to initiate and sustain operations in both contingency and training environments. Establishing an organization tasked with these responsibilities would enhance enterprise-level ACS planning, execution, monitoring, and control and thereby improve Air Force and Joint C2.

Figure 2.3
TO-BE Vision—A Single Organization to Integrate Assessments and Direct Actions to Balance Support

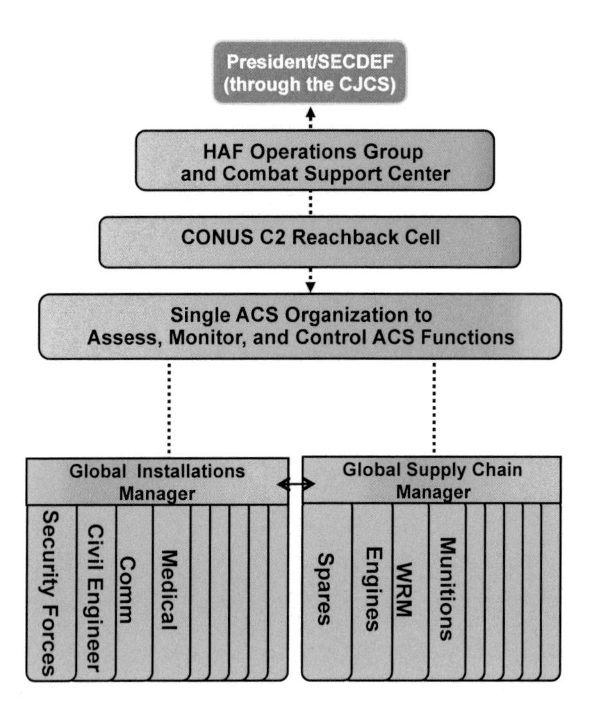

NOTE: Processes we defined in this analysis that are not currently assigned to an organization are shown in orange. We do not show ACS functions that fall outside the installation support and supply chain grouping (such as chaplain, historian, acquisition, and test and evaluation) although they also need to be globally managed and integrated to provide a complete picture of ACS capabilities and constraints.

The single organization responsible for ACS C2 presented in the vision would include a supply chain manager and an installations support manager, as well as other ACS functions that fall outside those groupings. The supply chain manager would be responsible for conducting supply chain assessments, configuring supply chains to meet operational needs, and developing supply chain mitigation strategies. Supply chain management would include directing and monitoring the performance of the repair network. The repair network may be organized under separate management (represented by the vertical green boxes in Figure 2.3), but it should be integrated in the overall supply chain (as part of the green and orange horizontal box in Figure 2.3) to best meet operational requirements. During steady state, supply chain management would manage the day-to-day, in-garrison supply chains needed to support organize, train, and equip

(OT&E) responsibilities. The installations support manager would be responsible for maintaining the home-station installations support needed to meet operational OT&E needs and for developing deployable packages needed to open and sustain FOLs. This manager would also be responsible for balancing ACS installations functions.[9] Other ACS functions, such as the chaplain and acquisition, may fall outside installation support and supply chain management, but they should also be managed from an enterprise perspective and integrated with other ACS capabilities and constraints.

This ACS integrating organization could also include reachback support to forward C-NAF and component major command (C-MAJCOM) staff to evaluate the supportability of different options for combining resources to achieve specified objectives as defined by the CCDR (shown in orange as the "CONUS C2 Reachback Cell" in Figure 2.3). Establishing the ability to perform risk assessments within the short decision cycles required by military leadership would entail investments in modeling capabilities and staff development.

In past analyses, RAND has called the organization responsible for conducting and balancing integrated assessments, including Air Force forces (AFFOR) reachback support, a *global integration center* (GIC) (see Figure 2.4). During a recent experiment organized by the Air Force Command and Control Integration Center (AFC2IC) as part of the Joint Expeditionary Force Experiment (JEFX), a portion the GIC concept was demonstrated. The Agile Logistics EXperiment (ALEX) used the operational support facility (OSF) at the Ryan Center, Langley Air Force Base (AFB), Virginia, to bring together several stovepiped ACS resources and provide three C-NAF staffs information about the Air Force's enterprise ability to support their COAs, thus testing the AFFOR reachback piece of the GIC concept (indicated by the red box drawn in Figure 2.4).[10] Instead of AFFOR staffs reaching back to each individual ACS functional manager for stovepiped capability assessments, there was an ACS reachback cell in the Ryan Center (a C-NAF reachback organization) that provided an assessment of the ability to generate sorties, as well as the ability to open FOLs for a select number of supply chain and functional capabilities.[11] The ACS reachback cell within the OSF provided a centralized location where C-NAFs could find enterprise-wide ACS capabilities and constraints.

---

[9] The Global Base Support (GBS) initiative at AFMC is supposed to standardize many core base operating support functions like civil engineering and communications. Although not designed as an installation support manager, GBS could play a role in this function.

[10] See Appendix A for more information about ALEX.

[11] Spare parts and engines were assessed to determine sortie generation capability. Civil engineers, SF, communications, medical, and WRM were assessed to determine FOL capability.

Figure 2.4
Part of the GIC Construct Was Tested During ALEX

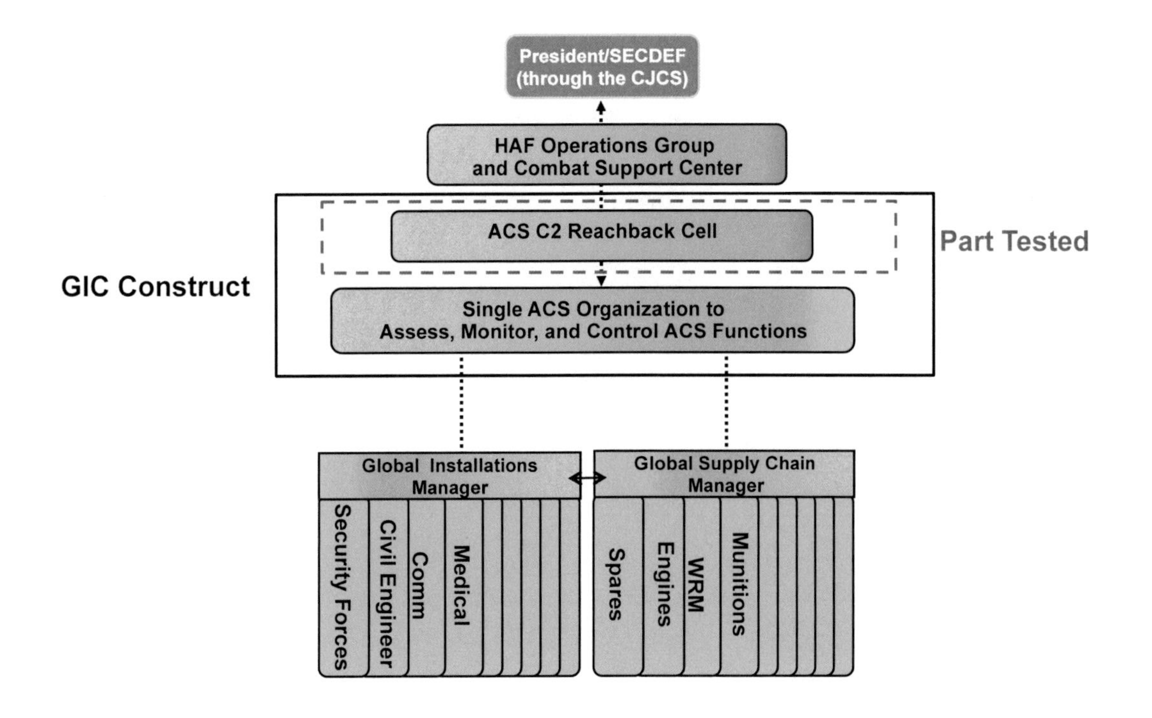

An ACS reachback cell within a C-NAF and C-MAJCOM reachback center may not be very costly. As demonstrated during ALEX, only a few people are needed to be on-site at the ACS reachback cell, as the majority of the analyses are conducted off-site and provided virtually.

During ALEX, the ACS reachback cell provided information to the C-NAFs and C-MAJCOMs for their planning processes; however, there was no organization with the authority to balance resources across competing demands or develop mitigation strategies. The TO-BE vision presented here goes further and calls for an organization responsible for assessing, monitoring, and controlling ACS activities across the enterprise.[12]

A single ACS organization with responsibility and direct management authority over all ACS functions could also ensure that each functional capability's manpower is aligned correctly during peacetime for wartime needs. In a recent RAND analysis, researchers

---

[12] Although ALEX only tested the reachback assessment part of the vision presented here, it was considered a success. Based on that success, organizers and participants conducted a second experiment (ALEX II) in August 2011 to further develop the concepts and implementation strategies.

found that some ACS manpower could be realigned to meet future OSD plans more effectively and efficiently.[13] The research team found that if manpower realignments were allowed, end strength could be reduced while still meeting OSD scenario requirements, generating a net savings of several hundred million dollars per year from manpower costs. Further, steady-state deploy-to-dwell ratios and home-station workloads during deployments could be improved for many career fields.

However, having a single ACS organization with responsibility and direct management authority over all ACS functions could present span-of-control issues. There may also be some risk in having a single point of failure without backup or redundancy built into the system.

A second, less-centralized option is presented in Figure 2.5. Instead of a single ACS organization, there could be separate supply chain and installation support organizations that are then brought together, integrated, and balanced at an integration cell, perhaps at the C2 operational support center (OSC). The other ACS functional capabilities outside of installation support and supply chain management should also be globally managed and integrated and balanced at the integration cell. The ACS reachback organization could either be part of the integration cell or it could be separate.

---

[13] Most ACS career fields derive manpower requirements from home-station installation needs, not expeditionary demands. This creates inherent imbalances for ACS manpower relative to expeditionary requirements: more military manpower in some areas than the Air Force could conceivably need and much less manpower in other areas than the Air Force would need to execute future OSD plans. If manpower within the active duty and reserve component were realigned, these imbalances could be remedied. The realigned ACS manpower mix would better meet surge and steady-state operations at the same or reduced end strength. See Patrick Mills et al., *Balancing Agile Combat Support Manpower to Better Meet the Future Security Environment*, Santa Monica, Calif.: RAND Corporation, RR-337-AF, 2014.

Figure 2.5
Organizational Option 2—Separate Supply Chain and Installation Support Capabilities Brought Together at an Integration Cell

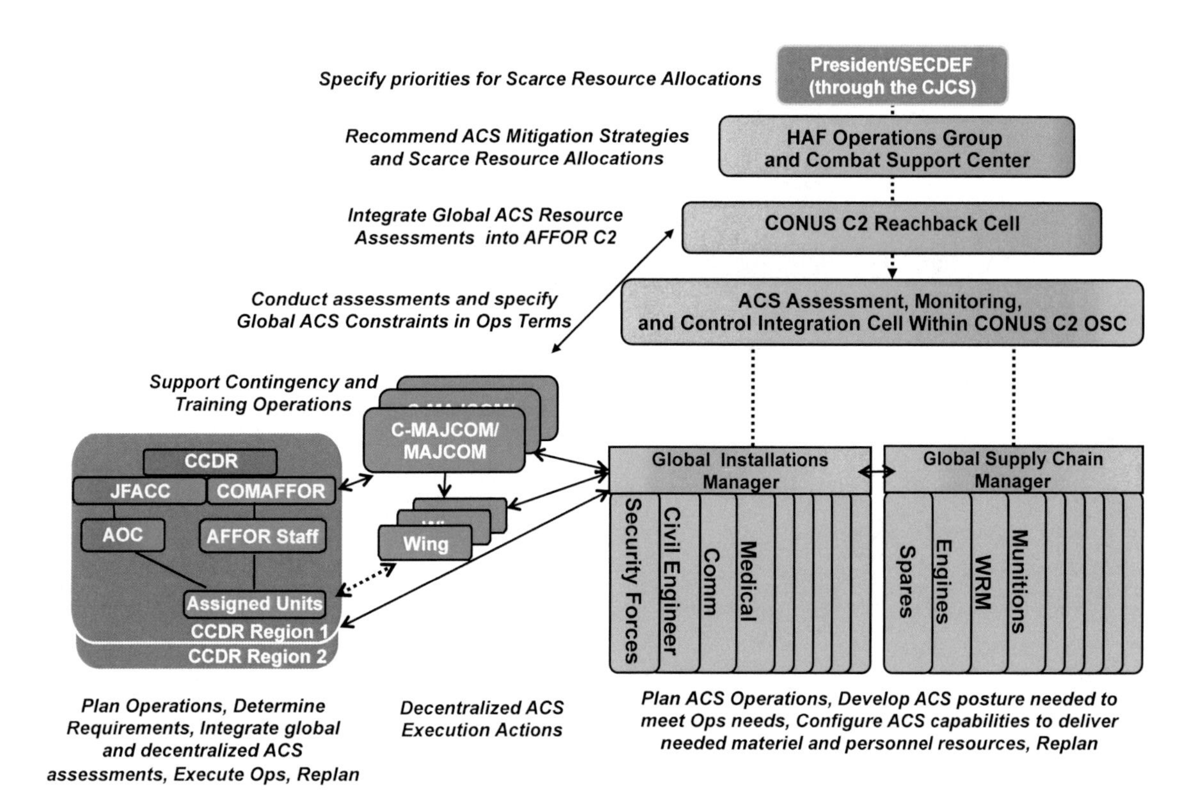

This second option may present less span-of-control issues, but it introduces another level of coordination before allocation and priority decisionmaking can occur.

Another option, a decentralized approach, was the organizational option used during ALEX (shown in Figure 2.6). In this option, individual resources are managed and assessed separately. The results are then fed into an integration cell that integrates the stovepiped analyses. Again, the ACS reachback cell could be part of the integration cell or it could be a separate organization. This is the organizational option that is partly in place today.

Figure 2.6
Organizational Option 3—Decentralized Approach

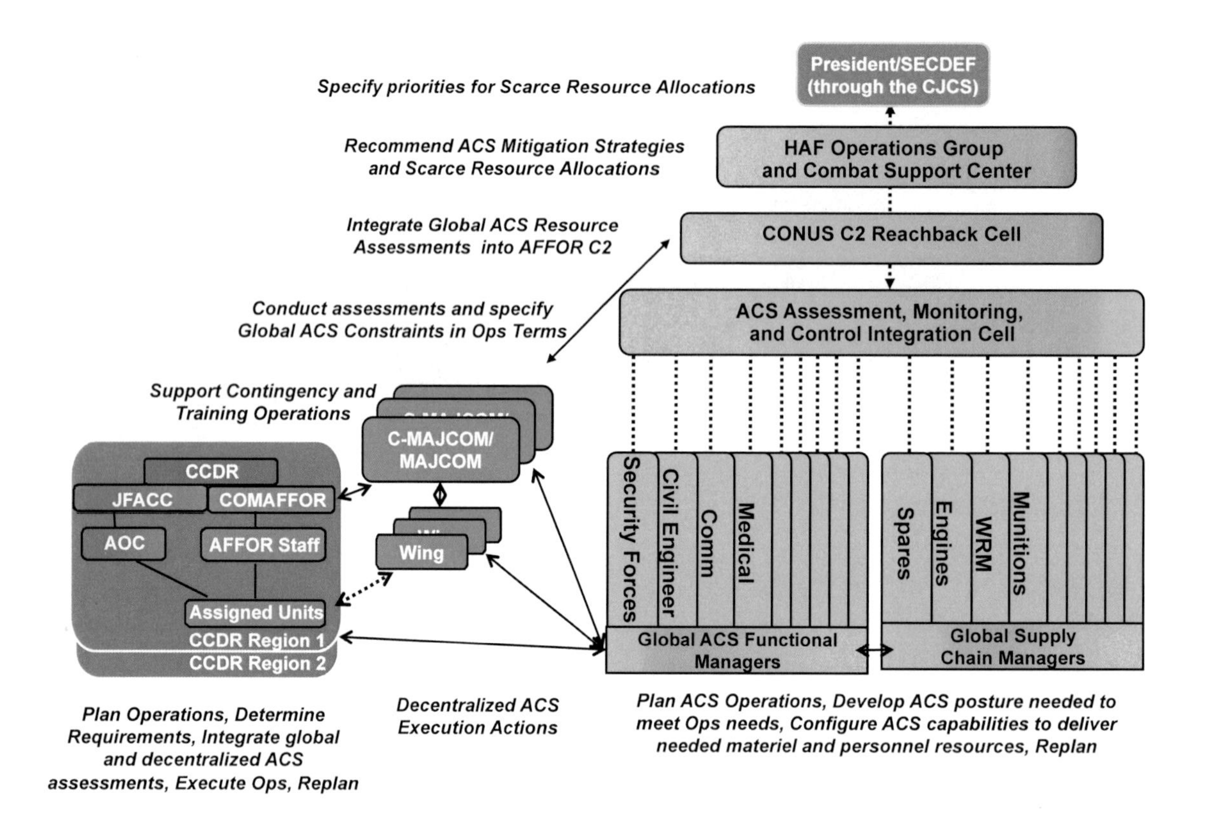

There were two problems experienced with organizational option 3 during ALEX. First, there were no standard business rules for the individual stovepiped resource analyses. Each resource analysis was completed separately with different rules and assumptions. What was considered green (or good) in one functional capability may have been considered yellow (or marginal) in another. Secondly, there was no organization with the authority to balance resources across competing demands or direct the development of mitigation strategies. In this construct, if there are competing demands, the issue must be raised to the AF/CV level for decisionmaking.

If organizational option 2 or 3 is implemented, the Air Force will need to determine where ACS reachback support would reside. ACC, as the force provider and the lead integrator for the C2 core capability, could serve as the assembly point for the capability assessments. The OSF at the Ryan Center, Langley AFB, Virginia, as exercised during ALEX, is one option. The OSF was designed to provide C-NAF reachback support to both the AOC and AFFOR staff as outlined in PAD 10-02. Another option is to have the integration occur at a supporting command.

Regardless of the organizational construct used, the process of bringing together, balancing, and presenting an integrated view of Air Force capability and constraints is vital to enhancing Air Force and Joint C2 processes. Today, combat support is treated as a set of unrelated resources, making it difficult for the AFFOR staff or global resource managers to produce timely integrated capability assessments. A single organization tasked with global ACS responsibility could help C-NAF and C-MAJCOM staffs identify the most binding ACS constraints and develop mitigation strategies across resource supply chains and functional capabilities, enabling the C-NAFs and C-MAJCOMs to help the CCDRs manage the risk associated with their contingency plans (CONPLANs).

Arbitrating Between and Among Competing Demands

Once binding constraints and mitigation strategies are identified, there should be a process to arbitrate across competing demands. As stated previously, there are not enough ACS assets to satisfy every operational demand as outlined in OSD planning guidance. A defined process is lacking for determining which operation will have priority and which planners will need to replan because assets are being reallocated to another theater or because they will not have all the assets they planned for. The Air Force has not formally designated an organization to seek priorities from the President and SECDEF for allocating scarce resources among AORs. Currently, planners operate under the assumption that sufficient ACS resources exist and will be allocated to them when needed. However, this will not be possible if there are simultaneous, nearly simultaneous, or increased and continued steady-state events. Scarce resources and increased demands necessitate the development of prioritization processes.

The ALEX experiment, hosted by the AFC2IC, demonstrated this necessity. During ALEX, several plans were evaluated—one contingency and two operation plans (OPLANs). The analysis showed the impact of allocating scarce resources among AORs. Each plan was evaluated independently, and then the plans were evaluated again simultaneously.[14] As expected, there were not enough ACS resources to support all three operations simultaneously. However, the ACS reachback cell at the OSF was able to quantify the capability shortfalls and relay that information to the C-NAF and C-MAJCOM staffs. This helped the C-NAF and C-MAJCOM staffs understand the shortfalls and risks associated with their plans and allowed them to replan their COAs as necessary. Missing in the experiment was a process to determine which plan would have priority and how that priority would be communicated to the C-NAF and C-MAJCOM staffs and to global supply chain and functional capability managers. Since no process is

---

[14] We developed briefing charts (not available to the general public) showing the experiment results.

currently defined, the exercise participants in the OSF explored several different prioritization scenarios. The impacts of each were relayed to the C-NAF participants.

An example of the type of analysis conducted in ALEX is shown in Figures 2.7, 2.8, and 2.9. This is a notional example of how resources could be allocated and reallocated across AORs to meet designated priorities. Figure 2.7 shows notional data for five independent ACS resources—civil engineers (CEs), SF, communications, medical, and contracting—across several AORs. Integrated together, these five independent capabilities could determine the ability to establish an FOL at a given location.

Figure 2.7
Notional Assessment of Ability to Establish FOLs

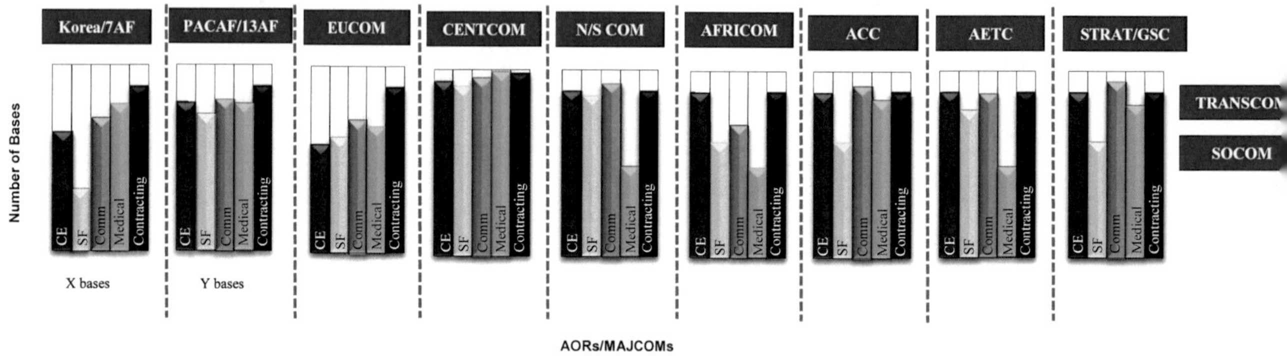

In the notional example shown in Figure 2.7, the ability to open FOLs in Korea is constrained by SF capability. If Korea was given priority over other plans, SF capabilities could be reallocated from other AORs to increase FOL capability in Korea. Figure 2.8 illustrates how FOL capabilities can be increased in Korea by reallocating SF capabilities from the Pacific Air Forces (PACAF). However, there is a cost in SF capability to PACAF if capabilities are reallocated elsewhere.

Figure 2.8
Notional FOL Capability Can Be Improved by Reallocating from Another AOR

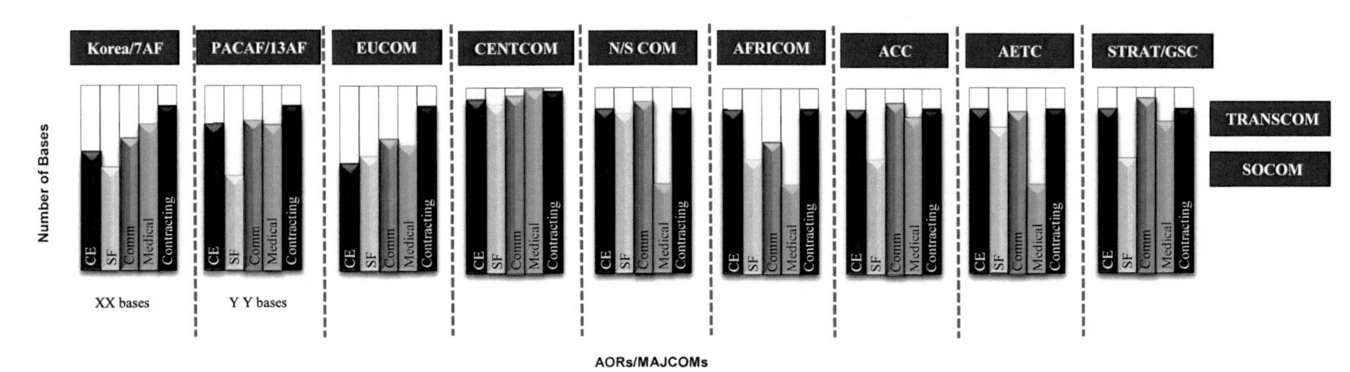

If desired, Korean FOL capability could be further increased by reallocating assets worldwide in all five resources areas (see Figure 2.9). Again, there is a capability cost to the other AORs.

Figure 2.9
Notional FOL Capability if All Resources Are Reallocated

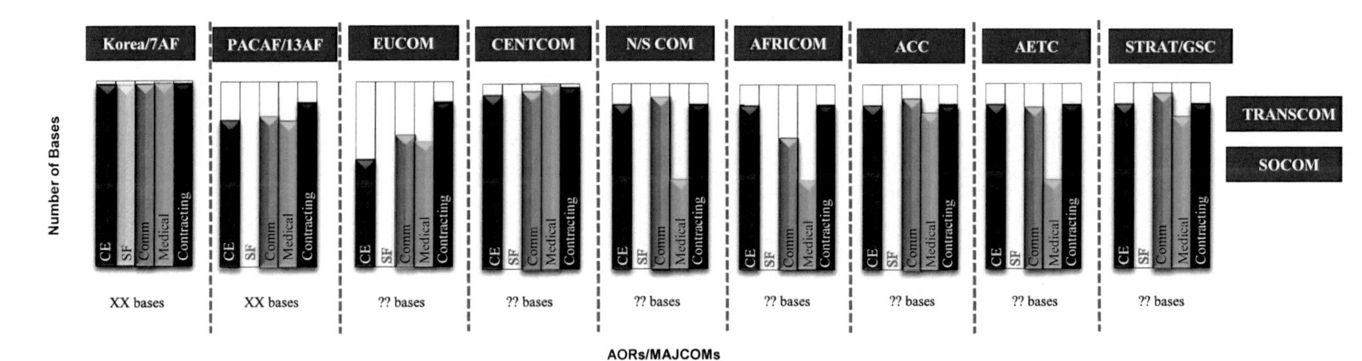

There should be a defined process for assessing risks and allocating scarce resources according to the President's and SECDEF's priorities. Without one, each AOR is operating in isolation, assuming they will receive assets when an operation commences in their AOR.

Process Summary

The Air Force has taken steps to address pieces of the three main process shortfalls outlined above by managing a few functional capabilities from an enterprise perspective and by establishing the AFGLSC (now part of the Air Force Sustainment Center), which could be used to bring together several classes of materiel. However, there is much work still needed. The issues are larger than any one Air Force organization. They cross many lines of authority and responsibility. Without designating a single ACS organization with the responsibility of addressing the issues and the authority to make changes, these process gaps may persist. A single organization and commander needs to be identified and given the authority to move the Air Force toward an integrated C2 vision. However, this transition to a single ACS authority would take time to implement. In the meantime, there are many other improvements that can be made quickly, such as improving doctrine, guidance, and instructions.

Doctrine, Guidance, and Instructions

Codifying the role of ACS in C2 in doctrine, guidance, and instructions is imperative for long-term implementation. It creates standard, repeatable processes to plan, execute, monitor, and control ACS enterprise actions to achieve specific and supportable operational needs. It also enables COMAFFOR and major command (MAJCOM) staffs to concentrate on developing supportable plans for meeting contingency and training requirements, while the ACS enterprise concentrates on delivering needed resources.

While progress has been made in demonstrating how ACS planning, execution, monitoring, and control concepts enhance Air Force and Joint C2 (during the ALEX experiment), there is still much work to be done to capture the processes in doctrine, guidance, and instructions. Once Air Force–level guidance defines roles and responsibilities, MAJCOMs can create and modify their instructions. For example, Air Force Materiel Command (AFMC) Supplement 1 to Air Force Instruction 10-401 (AFI 10-401 AFMCSUP I) provides basic direction:

> The AFMC Operational Plans Division (AFMC/A3X) is the focal point for coordinating all plans (whether produced by Headquarters AFMC or other AFMC entities) with other MAJCOMs. The AFMC Exercise Program (AFMC/A3X), in conjunction with AFMC inspector general (IG) and other OPR [office of primary responsibility] functions in direct reporting units (DRUs), will coordinate an annual process whereby all AFMC-scheduled exercises and AFMC-scheduled exercise–related activities (readiness exercises, IG inspections involving events termed exercises, experiments, wargames, demonstrations (capabilities/technology/other) termed exercises, etc.) are—to the maximum extent practical—synchronized within the Command and with the AEF [Air and Space Expeditionary Force] battle rhythm. The overall process is detailed in Air Force Materiel Command Instruction (AFMCI) 10-204, AFMC Exercise-Related Activities and Support.
>
> … Headquarters AFMC A-Staff/Functional Directorates develop, write, and update the annex or appendix [in AFI 10-401 AFMCSUP I] detailing their functional support to each AFMC war and crisis action plan. Each AFMC plan summary must include instructions for implementing the planned action (checklist, plans, or other procedure) and level of

> Command responsible for preparing the implementing document. The Headquarters AFMC Staff[15] reviews installation-level plans to ensure consistency and adequacy in supporting AFMC/Air Force plans. Headquarters AFMC/A3X provides the results of the Staff analysis to the originating installation for proper action.[16]

AFMC established the AFMC Wargaming Integration Office (AFMC/A8XI) as the Command's lead organization for participation in Title X wargames. The major focus areas of the office are to provide some logistics realism for the games, ensure results and feedback are integrated into AFMC ACS planning and programming cycles, and channel findings and results back to the respective agencies for their internal action. AFMC/A8XI personnel are also participating in selected exercises and experiments. All of these actions will provide an improved AFMC focus on contingency operations.

However, detailed roles and responsibilities are not included in doctrine, guidance, and instructions. Specifically, there are no AFMC instructions that address

- Headquarters AFMC's role in OPLAN and contingency planning (including LSAs), COCOM exercises, wargames, and experiments
- how each AFMC A-staff, center, and directorate supports OPLAN or other major contingency planning (including LSAs), COCOM exercises, wargames, and experiments
- the products that each organization should produce in support of one of these events
- command organizations' responsibilities and relationships with outside organizations, such as C-NAF/CCs, C-MAJOM/CCs, Air Staff, and the AFC2IC, or the relationships between each AFMC organization, such as the Air Force Sustainment Center's relationship with the Air Logistics Complexes (ALCs) within the new AFMC center construct
- the Air Force Sustainment Center's (which includes the former AFGLSC's) responsibility to conduct proactive OPLAN LSAs and strategies to compensate for spare shortages written in Surge Contingency Plan 70s and each ALC's Plan 70 accordingly and provide the results to the appropriate C-NAF
- the required training curriculum for all personnel who may participate in these events
- the inspection criteria and command inspection schedule for each command organization to support these activities.[17]

---

[15] Or, under the AFMC reorganization, as part of the new AFMC Sustainment Center.

[16] U.S. Air Force Materiel Command, Air Force Materiel Command Supplement 1 to Air Force Instruction 10-401 (AFI 10-401 AFMCSUP I), *Air Force Operations Planning and Execution*, July 29, 2010a, Chapter 11, "Roles and Responsibilities."

[17] We have documented specific changes necessary to some instructions in Appendix B of this publication.

Additionally, AFMCI 10-204, while still in draft form, appears to be too general in nature and does not specify the AFMC C2 nodes required for assessments of global resources and process performance.[18]

AFI 10-401 also needs to be updated.[19] One specific responsibility of Air Force supporting commands is to complete and submit LSA results to supported COMAFFORs. This instruction should be updated to include requirements for global ACS integrated capability assessments and prioritization rules for allocating scarce resources among C-NAFs and C-MAJCOMs. It should also include the Air Force Sustainment Center's (which includes the former AFGLSC's) role in LSAs as part of crisis action and contingency planning. Currently, there is no mention of the Air Force Sustainment Center in AFI 10-401.

Additional instructions could be written mandating an ACS reachback cell within the OSF to support C-NAF and C-MAJCOM AFFOR staffs and AOC reachback operations as demonstrated in the ALEX experiment. Air Force PAD 10-02 supports C-NAF reachback functions, including ACS global resource assessments.[20] It directs that C-NAFs be manned day-to-day to respond to Phase 0 and Phase 1 operations and maintain readiness to support other phases and that C-NAFs no longer man to the 72-hour surge requirement. This PAD also directs the test and development of an OSF capability, which is fundamental to the future of effective and sustainable C2 for geographic C-NAFs and C-MAJCOMs. The PAD states that an OSF should include the following:

- C-NAFs' and C-MAJCOMs' core continuity of operations (COOP) for those C-NAFs that are conducting ongoing regional combat operations
- the primary reachback facility for AOC and AFFOR staff, potentially eliminating some of the requirements for augmentation
- a capability for training, exercise, and experimentation support for AOC and AFFOR system capability.

If the Air Force decides to use the OSF for global integrated ACS capability assessments for reachback, an instruction detailing the OSF's specific responsibilities is required. With a modest number of assigned personnel and using virtual support from some organizations (for example, ACS supply chain and functional capability managers), the OSF could perform global ACS assessments and provide feedback to C-NAFs and C-

---

[18] U.S. Air Force Materiel Command, Air Force Materiel Command Instruction 10-204, *Exercise Program*, August 31, 2010b.

[19] U.S. Air Force, Air Force Instruction 10-401, *Air Force Operations Planning and Execution*, December 7, 2006c.

[20] U.S. Air Force, 2010b.

MAJCOMs. Being an entirely new concept, multiple MAJCOM (for example, AFMC and ACC), C-NAF and C-MAJCOM, and Air Force instructions require additions and modifications.[21]

AFMC's role in planning, exercises, and experiments is critical for assessing how well the ACS enterprise can support near-term conflicts with existing resources. Air Force and AFMC publications should be updated to reflect changes and new processes to help institutionalize AFMC's role in ACS planning, execution, monitoring, and control.

Training and Career Development

Trained personnel are necessary to help remedy the shortfalls in the process outlined above—conducting integrated and balanced capability assessments and developing scarce resource allocation schemes. It may be necessary to develop new ACS planning, execution, monitoring, and control curricula, which could then be incorporated into existing or new training courses. Enhanced ACS curricula should train on topics such as combat support doctrine, policy, and guidance; AFFOR staff and AOC combat support processes; integrated ACS capability assessments; operationally relevant ACS metrics; and new decision support tools, as they are developed. Expanded training could include the testing of new tools, systems, and processes before they are fielded.

New and enhanced curricula could be incorporated in training courses such as the joint services introductory course for basic AOC processes, the Contingency Warplanning Course (Maxwell AFB), the C-NAF Commander's Course (Maxwell AFB), and courses at the Air War College and Air Command and Staff College, as well as courses taught at the Air Force Institute of Technology and other civilian universities that have supply chain curricula.

Career-path planning for combat support personnel might include assignment to warfighting command-level positions in supply, transportation, logistics plans, civil engineering, or services, with the intent of creating senior combat support personnel with the skills needed to fill AFFOR A4 (logistics) and A7 (installations and mission support) staff and COCOM joint staff ACS positions. Those combat support officers with a strong C2 background can be groomed for leadership positions. Additional education and training might be needed for those who will occupy key ACS assignments responsible for integrating combat support into the joint system, such as in the COCOM J4 staff, the COMAFFOR A4/7 staff, and in the AOC. While the number of these positions is not large, they are key to the development of feasibly operational plans.

---

[21] Specific inputs for PAD 10-02 can be found in Appendix B of this report.

Finally, the Air Force should ensure that operators are trained to create operational planning teams (OPTs) (understanding their uncertain planning environment) that include combat support planners in a timely manner. Operators should understand what combat support planners need and when, and combat support planners should understand the limitations and uncertainties within which the operators work. Processes that define how operations and combat support planners should work together need to be codified in guidance and instructions and routinely exercised so these processes become institutionalized. Only by training both groups to understand both sides of the planning equation and communicate effectively will this link between operational and combat support planning be forged and sustained.

Training and realistic exercises are critical aspects of the link between combat support and operational planning. Educating both combat support and operations personnel about their roles in the context of campaign planning will enable more-effective communication and facilitate the integrated decisionmaking process outlined in the operational architecture.

Gap and Shortfall Summary

While progress has been made in improving ACS planning, execution, monitoring, and control processes through establishing some global managers and the C-NAFs, transforming the logistics enterprise, and designating core functions,[22] there are still improvement actions that need to be taken. Global supply chain and functional capability managers are being established to manage and control some resources, but other resources are managed theater by theater, without an enterprise view of Air Force capability. No organization is charged with integrating and balancing stovepiped resource assessments to provide capability and constraint information to the warfighter. Nor has the process by which to allocate scarce resources across competing demands been defined and written into doctrine, guidance, and instructions. Tools and systems to help analyze Air Force ACS capability and limitations are limited. Each of these gaps underscores the need for standardized, integrated ACS processes focused on operationally relevant results codified in doctrine, guidance, and instructions and led by an organization charged with the responsibility and given the authority to manage ACS capabilities.

---

[22] For more information about recent Air Force process improvements, see Chapters One and Two of Lynch et al. (2014).

3. Conclusions and Recommendations

The focus of this analysis is on how enhanced ACS processes can be implemented and integrated within the Air Force and Joint C2 enterprise. The updated architecture, developed as a companion piece to this analysis, provides the vision for enhanced C2.[1] We use this architecture to identify specific improvements that are needed at the strategic and operational levels. We evaluate C2 nodes from the level of the President and SECDEF to the units and sources of supply. We also evaluate these nodes across operational phases—from readiness preparation through planning, deployment, employment, sustainment, and reconstitution. The concepts we describe have been widely vetted with senior operational and ACS leaders, and there is agreement that enhanced processes and tools and better-trained personnel are needed to integrate enterprise-wide ACS capability assessments for COA development, time-phased force and deployment data (TPFDD) analysis, and other C2 activities. The Air Force has acted to include enhanced ACS processes in OPLAN development (AFGLSC LSA analysis), exercises, and experiments (such as ALEX); however, there is more work to be done.

In the process area, the overarching shortfall is the inability to provide an enterprise assessment of combat support capability and constraints to inform trade-off decisions so that scarce resources can be effectively and efficiently used to meet Air Force operational priorities. To address this shortfall, processes need to be enhanced on several different levels:

- within individual ACS supply chains and functional capabilities: Global capabilities need to be assessed in a standard, repeatable manner that links directly to the ability to meet operational requirements.
- across individual supply chains and functional capabilities: Individual supply chain and functional capability assessments need to be integrated and balanced into a set of capabilities that can be used in planning and replanning processes.
- within the Air Force: There should be a defined process to arbitrate between and among competing operational demands.

Further, ACS processes are not established and defined in doctrine, guidance, and instructions; tools and systems are lacking; and there is no identified authority to direct and redirect resources across the enterprise.

---

[1] Lynch et al. (2012) describes in detail a strategic- and operational-level C2 architecture integrating enhanced ACS processes.

Recommendations

While there is general agreement on the need for and value of the processes we describe, there are differing views as to who should be responsible for developing them. For instance, the C2 Core Function Team, which is responsible for developing C-NAF and C-MAJCOM C2, views the integration of information coming from the enhanced ACS processes into the C-NAF and C-MAJCOM node for use in operational C2 as its programming support responsibility. However, it is the responsibility of the ACS Core Function Team to program for developing and sustaining decision support systems to produce that information.

One way for the Air Force to begin addressing these challenges is to call a C2 symposium to bring together Air Force communities that play key roles in this area. The symposium could be used to define assessment and control technique requirements; vet capabilities; identify necessary changes in doctrine, guidance, and instructions; develop needed training enhancements; and define organizational roles and responsibilities.

The architecture developed as a companion to this analysis can be of particular importance if enhanced ACS processes are developed separately by several organizations (such as AFMC, AETC, and ACC) and the information from these processes is integrated at C2 nodes at the operational and strategic levels by other organizations (such as ACC and the Air Staff). The architecture identifies the processes that can be developed by one organization and the outputs that can be integrated into other processes developed by other organizations. In short, it does not matter who develops the processes and associated systems from an architecture perspective. The architecture can be useful for ensuring the information coming from one system is integrated into the other systems by showing the system of systems perspective.

While progress has been made in improving ACS planning, execution, monitoring, and control processes, there are still improvement actions that need to be taken. The issues are broader than any one Air Force organization, thus closing these gaps may be difficult without designating a single leader as the ACS planning, execution, monitoring, and control commander. A single ACS organization should be given the responsibility of and the authority to address the gaps outlined above.

Moving to a single ACS authority would be a large cultural and organizational shift for the Air Force that would take time to implement. In the meantime, there are many actions that can be taken to improve ACS planning, execution, monitoring, and control. Specifically, to institutionalize and further enhance the integrated assessment and allocation processes discussed in this work, the Air Force should continue to experiment and participate in exercises that can demonstrate and enhance these processes over time. For example, we recommend that the Air Force continue the ALEX series of experiments using an ACS reachback cell at an OSF to conduct global assessments and provide capability and constraint feedback to C-NAF and C-MAJCOM planners. The ACS reachback cell could also provide analysis to planners during exercises such as Austere Challenge, Terminal Fury, and Ulchi Freedom Guardian. Key to these assessments is an organization with tools and trained personnel tasked with the responsibility of supporting C-NAF and C-MAJCOM planning, exercises, and experiments.

The enhanced ACS processes the Air Force implements need to be codified in doctrine, guidance, and instructions. The roles and responsibilities of each C2 node, including logistics, operations, and installation staff; Air Force commanders; MAJCOMs, specifically AFMC; and others, such as an ACS reachback cell, should be delineated in doctrine, guidance, and instructions. Specifically, the LSA process defined in AFI 10-401 should be updated to require global integrated ACS resources assessments and prioritization rules for allocating scarce resources.[2]

Once Air Force–level guidance defines and assigns roles and responsibilities, AFMC and the Air Force Sustainment Center (including the ALCs) need to develop corresponding instructions outlining command organizational roles in C2 processes, such as

- the command role in proactive OPLAN assessments and contingency planning (modify AFMCI 10-204)
- the Air Force Sustainment Center's (which includes the former AFGLSC's) and Headquarters' roles in developing proactive risk mitigation strategies (modify Headquarters and ALC OPLAN 70s)
- a single point of contact (POC) to direct guidance for AFMC across OPLANs, contingency operations, exercises, experiments, and wargame C2 responsibilities.

Without clear guidance, enhanced ACS processes may not become institutionalized in how the Air Force does business. Advancements and enhancements in ACS could be lost without clear directives providing roles, responsibilities, and authorities.

The Air Force has moved forward in achieving the vision presented in the operational architecture, but our research shows that many actions still can be taken to improve ACS planning, execution, monitoring, and control processes; doctrine, guidance, and instructions; training and career management; and tools and systems.

---

[2] U.S. Air Force, 2006c.

Appendix A. The Agile Logistics Evaluation EXperiment

During JEFX 11-1 in January 2011, the AFC2IC conducted an experiment called the Agile Logistics Evaluation EXperiment—the name has now been shortened to the Agile Logistics EXperiment (ALEX). During ALEX, the AFC2IC stood up an ACS cell in the OSF at the Ryan Center (Langley AFB, Virginia) to conduct ACS assessments for existing OPLANs. These ACS assessments were identified as a process shortfall or gap in the analysis documented in this report. As part of the DOTMLPF analysis conducted with the architecture, we identified the need for (1) the integrated analyses and (2) an organizational construct to be charged with the responsibility for conducting the analyses—what we call a *global integration center* (GIC).[1] ALEX demonstrated a portion of the GIC concept using a small reachback cell to perform ACS assessments to identify ACS capabilities and constraints. The same GIC concept was tested again with expanded capabilities in ALEX II in August 2011.

Background and Motivation

As documented in previous RAND analyses, warfighters today are developing contingency COAs for their AORs with limited information about global ACS resource availabilities.[2] Essentially, they operate under the assumption that sufficient ACS resources exist and will be available to meet their priorities. However, for a number of reasons, including budgetary constraints, the inability to perfectly predict demands, and variability in supply processes, there will always be imbalances between the ACS resources that are available and those required to meet operational demands.

Some resources are managed globally and resource capability assessments are available to the warfighter. For example, the Air Force Sustainment Center (which includes the former AFGLSC) manages the worldwide supply of spare parts, and the GACP manages munitions globally. However, these capability assessments are stovepiped and similar analyses are not available for all capabilities. Individual stovepiped resource capabilities need to be integrated with other categories of support (such as materiel, personnel, infrastructure, and transportation) to provide insights on how to allocate scarce resources among competing demands (see Figure A.1). Currently

[1] For more information on process shortfalls and gaps, see Chapter Two.

[2] Robert S. Tripp et al., 2012.

no organization is appointed to conduct integrated capability assessments. Ultimately, the goal should be to determine how alternative resource allocations affect bombs on target or other desired effects.

Figure A.1
No Organization Currently Provides Integrated ACS Capability Assessments

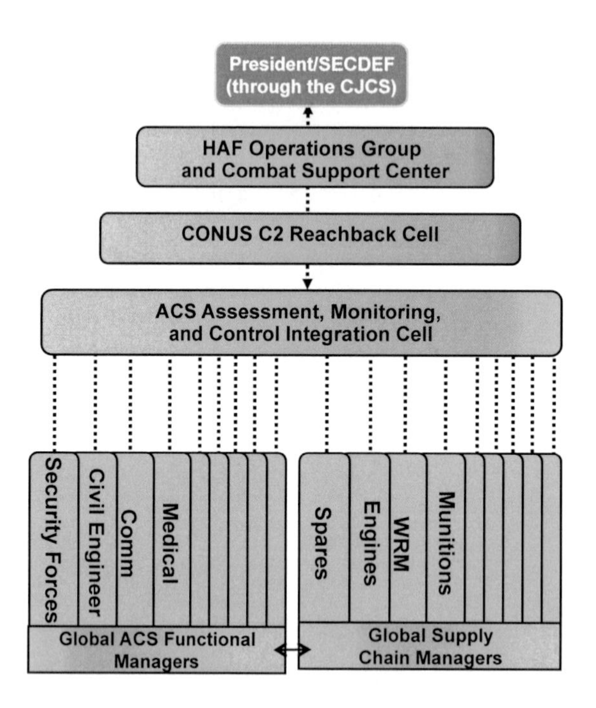

There is a range of options for how to assign responsibility for integrated assessments across ACS capabilities.[3] During ALEX, a decentralized approach was used. An ACS reachback cell in the OSF conducted assessments, identified the most-binding constraints, provided assessments to higher authorities, and executed resource allocation strategies when notified by higher authorities for a small set of resources. However, the ACS cell did not manage or control ACS resources, as recommended in the vision for an ACS integration organization. We discuss the experiment in more detail below.

[3] The vision presented in this analysis calls for a single ACS organization responsible for integrating and balancing ACS resources across the enterprise. However, there are other organization options presented in Chapter Two.

ALEX Objectives and Results

ALEX was designed to test portions of the GIC concept—for instance, could a global capability analysis be completed by an ACS reachback cell in time to inform C-NAF planning processes? The desired output at the end of the experiment was a timely and accurate assessment of OPLAN supportability for a specific group of resources.[4]

The first ALEX, conducted in January 2011, was able to produce the desired assessments for spare parts for F-15C/Es; engines for the F-15E; and FOL capabilities in several distinct UTC categories—civil engineers, SF, medical, communications, and basic expeditionary airfield resources (BEAR). The ACS reachback cell analyzed spare parts, engines, and UTC requirements for two separate OPLANs. The independent assessments were used to determine whether existing levels of resources and personnel could support each OPLAN independently without giving any consideration to the shortages of resources and personnel one plan may impose on the other. Then, both OPLANs were assessed simultaneously, with one plan having priority over the other (that is, the priority plan was given all requested personnel and resources available). The analysis determined the residual impact on the lower-priority plan. The outputs or products were analyses of logistic supportability, the ability to meet operational requirements, and any constraints or shortfalls CCDRs could expect for these OPLANs. The ACS reachback cell was able to show the impacts of allocating resources to one priority over another.

ALEX Methodology

To complete the analyses for ALEX, the AFGLSC used the PC-Aircraft Sustainment Model to download current worldwide spares and engine data. They then showed enterprise constraints and how those constraints impact aircraft availability and sortie generation capability. According to the survey results after the ALEX event, most people found the spares and engines analyses useful for planning.[5]

RAND analysts evaluated the global ability to support FOL requirements using FOL Assessment Model (FAM) and current AEF Reporting Tool (ART) data. Again, according to survey results, most users found the UTC FOL supportability analysis useful

---

[4] The details of ALEX 11-1 are located in U.S. Air Force, Air Combat Command, Air Force Command and Control Integration Center, Joint Expeditionary Force Experiment 2011, JEFX 11-1, *Agile Logistics Evaluation Experiment (ALEX) Final Report*, April 25, 2011.

[5] U.S. Air Force, Air Combat Command, Air Force Command and Control Integration Center, Joint Expeditionary Force Experiment 2011, JEFX 11-1, 2011.

for developing plans or replanning activities. We will discuss FAM in more detail in the next section.

FOL Assessment Model

FAM is a stand-alone Excel workbook that allows users to compare the supply of available UTCs (from ART) with a demand for resources (in a TPFDD).[6] ART provides a snapshot of UTC readiness. Each UTC is rated by its commander as either "Green," "Yellow," "Red," or "Incorrectly Postured." Also available in ART are the number of personnel currently deployed out of the UTC's full manpower complement and the UTC's P-code, which indicates whether it would generally be available during normal rotational operations or during the various levels of surge. The user has the option to select any of the ART availability levels when running FAM. The FAM user also has the option of drawing on any (or all) Active, Guard, and Reserve UTC pools, as well as the ability to allow employed-in-place units to count in the supply roster.

Using the user's inputs and run options, FAM tallies the availability of UTCs against the total requirements. FAM then reports the quantity and magnitude of UTC shortfalls, should any exist. FAM does not address sourcing decisions.

As an additional level of detail, FAM also reports shortfalls at the Air Force specialty code (AFSC) level. With this information, subject matter experts can evaluate whether a simple AFSC substitution might be sufficient to shift an unavailable UTC into the ready pool. For example, an available 3-level electrician might effectively stand in for a currently inaccessible 5-level electrician.

Summary

In today's resource-constrained environment, limited personnel and resources are available to support multiple operations. Command-level allocation and mitigation decisions are required. ALEX provided visibility on the global status of select ACS functions and their effects on both the higher- and lower-priority plans. Real-time, integrated supportability assessments were quickly available for C-NAFs to develop COAs, prioritize operations, and, if necessary, replan.

---

[6] TPFDDs are employment plans that itemize the requirement for UTC packages need to ship from home stations to FOLs.

ALEX validated the concept of using a central GIC for performing OPLAN or contingency supportability assessments in a distributed environment.[7] The GIC used a standard, repeatable process to plan, execute, and monitor OPLAN support activities.

---

[7] ALEX tested a decentralized approach to the concept in which data was brought together and presented from a single organization. It did not test management and control concepts as outlined in the vision (see Chapter Two).

Appendix B. Suggested Revisions to Air Force Documents to Enhance ACS Processes

In this appendix, we list the Air Force and AFMC publications that should be updated to codify the role of ACS in C2 processes. For each publication, we provide a synopsis of the relevant guidance as it pertains to ACS processes and point out where doctrine, guidance, and instructions may require changes to achieve the integrated vision for C2 in the future.

United States Air Force, Program Action Directive 10-02, *Implementation of the Chief of Staff of the Air Force Direction to Restructure Command and Control of Component Numbered Air Force*, June 2, 2010

This directive formalizes and implements CORONA South (Commanders Conference) Taskers and CORONA Top Taskers. Notably, it establishes the requirement for Deployment Orders (DEPORDs) for the Rapid Augmentation Team (RAT) and Air Reserve Component. The AFFOR C2 enabling concept, implemented by PAD 06-09, established a new Air Force component headquarters structure better able to support the CCDRs and provide C2 of AFFORs. Subsequently, the Chief of Staff, United States Air Force (CSAF) directed that C-NAFs be manned day to day to respond to Phase 0 and Phase 1 operations and maintain readiness to support other phases and directed that C-NAFs no longer man to the 72-hour surge requirement. This CSAF direction left C-NAFs with the risk of not being able to fulfill the full range of roles and responsibilities and, during a rapid transition to Phase 2 operations, could put CCDR's intent in jeopardy. To mitigate these risks, ACC led the implementation of a total force, Air National Guard (ANG) and Air Force Reserve Command (AFRC), 125-person RAT alignment plan for all phases of conflict.

Also of note is that PAD 10-02 directs the test and development of an OSF capability. It defines an OSF as a fundamental element for the future of effective and sustainable C2 for geographic C-NAFs and explains that an OSF should provide COOP to C-NAFs conducting ongoing regional combat operations; serve as the primary reachback facility for AOC and AFFOR staff, potentially eliminating some of the requirements for augmentation; and serve as a capability for training, exercise, and experimentation support for AOC and AFFOR system capability. The OSF will include an AN/USQ-163 Falconer AOC system capability.

It is essential that this PAD be modified to include the establishment of an ACS reachback cell within the OSF to support C-NAF AFFOR staff and AOC reachback operations as demonstrated in JEFX 11-1 (ALEX 11-1 and Austere Challenge 2011), as well as address the manpower requirements needed to maintain interfaces with C-NAFs, the Air Force Sustainment Center, global force management (GFM) functional capability managers, ACC/A4X, and others. We recommend standing up the Ryan Center with six personnel, possibly from the 710 Combat Operations Squadron (710 COS), to staff the ACS reachback cell at the OSF. The criteria for the six positions should include an analytic ACS background. Specific inputs to this directive could include:

- Section 2.3 or 2.4. In addition to describing the risk to C-NAF responsiveness and CCDR intent due to C-NAF manning, describe the need for global ACS enterprise assessments. Currently, C-NAF COA selections do not consider the ability of the ACS enterprise to support OPLANs and contingency operations. Therefore, C-NAFs are developing plans that may not be supportable from ACS perspective and are committing Air Force forces to operations with unknown and potentially serious support shortfalls. This was demonstrated in ALEX 11-1 when global ACS assessments were conducted on specific OPLANs. In this section, include language that addresses this and states that imbalances between needed global ACS resources and their availabilities require global assessments to ensure C-NAF COAs are supportable.
- Section 3.2.19.4. Expand to include testing the OSF ACS reachback cell's ability to conduct global ACS materiel and personnel assessments.
- Section 5.1.8. Include language that the OSF ACS reachback cell will be staffed by six members of 710 COS and stood up when CONPLANs or specific OPLAN assessments dictate (e.g., during LSA evaluations).
- Section 5.7.2. Add a third part to the OSF definition that describes the OSF as serving as a reachback entity to obtain global ACS assessments to determine if the global ACS enterprise has the capabilities to support C-NAF COAs.
- Section 5.9.1. Include language stating that AFFOR training includes an understanding that the OSF ACS reachback cell provides reachback capabilities to outline the types of global ACS resource assessments.
- Section 6.2.6.1.4. Indicate that, initially, six positions are required in the OSF to maintain interfaces with C-NAFs, the Air Force Sustainment Center (which includes the former AFGLSC), GFM functional capability managers, ACC/A4X, and others. These positions can come from the 710 COS and can be activated when periodic assessments are needed for OPLANs or when contingencies dictate. When activated, these personnel would be assigned to the OSF ACS reachback cell at the Ryan Center, Langley AFB, Virginia. The personnel in these positions can relieve and come from the RAT as described and need to have background and knowledge in the ACS fields that is analytic in nature (for

example, a Logistics Management Enlisted-to-AFIT graduate). Further explain the duty description of the full-time positions when activated within the OSF ACS reachback cell—facilitate ACS functional assessments; balance them based on the intractable personnel or resources; and then provide balanced assessments to C-NAFs, Air Staff, and ACS resource providers (for example, the Air Force Sustainment Center [which includes the former AFGLSC], ACC/A4X, and the civil engineering functional capability manager).

Air Force Materiel Command Instruction 10-204, *Exercise Program*, August 31, 2010 (Draft)

This AFMCI was in rough draft form when we reviewed it for this analysis. When completed, it will implement Air Force Policy Directive (AFPD) 10-2, *Readiness*, and AFI 10-204, *Participation in Joint and National Exercises*. It guides AFMC in its role as an Air Force MAJCOM when participating in Joint and national exercises and identifies AFMC exercise responsibilities; details the basic structure and objectives of the AFMC Exercise Program; establishes AFMC After-Action and Remedial Action Reporting; and describes funding for the AFMC Exercise Program. This instruction provides a cursory level of command exercise play, normally in support of joint or national exercises, to facilitate AFMC-led exercises with a focus on C2, communications, and responsiveness to crisis events.

As one of AFMC's capstone exercise instructions, AFMCI 10-204 is too general in nature and does not specify AFMC C2 nodes that should be tied to C-NAFs for the assessment of global resource levels and process performance. It does not address the specifics of how C2 should be exercised as part of global C2 enterprise. The instruction should define in detail how each AFMC A-staff, center, and directorate will demonstrate its support of full contingency planning for OPLAN or other major contingency planning and execution and as part of LSAs, exercises, experiments, and wargames. It should further define the C2 relationships between each AFMC organization (for example, the Air Force Sustainment Center's relationship with ALCs), as well as these organizations' relationships with C-NAFs and other participants (for example, the Air Force Sustainment Center's relationships with C-NAF/A3/5/4 and an OSC). The instruction should also address all metrics used to assess materiel support and supply chain effectiveness.

Participation in contingency operations, OPLAN LSAs, COCOM-sponsored exercises, selected experiments, and wargames are important for assessing how well the ACS enterprise can support current OPLANs and other near-term conflicts with existing resources. These events can improve existing and emerging Air Force and Joint C2 concepts by enhancing ACS contingency planning, execution, monitoring, and control processes and integrating these within the C2 enterprise. The lessons learned from each event can offer more realistic inputs and results in future events, providing a continuous improvement in expanding and enhancing ACS contingency planning and execution realism and readiness within Air Force and Joint C2 enterprise. AFMCI 10-204 should thoroughly detail the roles, responsibilities, and relationships of all AFMC directorates and centers in contingency operations, OPLAN LSAs, COCOM-sponsored exercises, selected experiments, and wargames; detail specific output products of all materiel supply chain C2 activities and link them to a specific schedule of events; detail each directorate and centers' responsibilities to C-NAF/CCs, Air Staff, etc.; define the required output products that should come from each event; and establish and detail the curriculum of a training workshop.

COCOM-sponsored exercises Terminal Fury and Ulchi Freedom Guardian are tied to major OPLANs and Austere Challenge is an annual joint exercise that tests U.S. European Command (EUCOM) and supporting commands' ability to plan and execute operations as a JTF Headquarters. All these exercises offer the opportunity to employ vital portions of the ACS contingency planning, execution, monitoring, and control processes; the commands go to war with current resources and force structure, and a major part of the supporting OPLANs is the LSA that is conducted to support them. AFMC should monitor and track combat support planning, execution, and control (C-SPEC) activities and tie them to the major OPLAN LSA process, then use the LSA results to test key portions of the ACS planning and execution process in the major COCOM-sponsored exercises. AFMCI 10-204 should document AFMC's participation in COCOM-sponsored exercises.

Participation in selected experiments tests the expansion and integration of new planning and execution capabilities, including new supply chain C2 processes and C-SPEC capabilities. AFMC participation will focus on testing ACS nodes and processes that are affected by AFMC, such as the ability to build/sustain needed FOLs given current and/or future resources; test C-SPEC, FAM, supply chain C2, and other expanded ACS planning and execution capabilities; and introduce and test files from other planning and execution systems. AFMC participation in these events will improve ACS realism and, when used in real-world contingencies, will offer more-accurate COA decisions to COCOMs.

AFMCI 10-204 should expand on the main focus areas of how all A-staff, directorates, and centers participate to provide C-NAFs with accurate ACS status when needed in contingency operations, OPLAN LSAs, COCOM-sponsored exercises, selected experiments, and wargames. This AFMCI should establish criteria for each A-staff, directorate, and center to meet and should direct periodic inspections to ensure compliance with this AFMCI and other AFIs. Finally, this AFMCI should direct and establish training criteria for key personnel in all A-staff, directorates, and centers.

It is essential that a single focal point be assigned, and AFMC Directorate of Air, Space, and Information Operations (AFMC/A3) would be a good choice. However, the instruction does not mention AFMC/A4, which should be included as an office of coordinating responsibility (OCR). Given AFMC/A4's large role with ALCs (and product centers, labs, etc.) and AFMC materiel sustainment policy issues, AFMC/A4 (logistics) is the equivalent of A3/5 (air, space, and information operations) in operational commands. Further, AFMC/A8/9 (strategic plans, programs, and analyses) should also be included as an OCR. AFMCI 10-204 should also capture AFMC/A8XI's role in selected exercises (as build-up to Title X wargame events). Because of the progress AFMC/A8X has made in establishing its wargaming office (A8XW) and its established participation in exercises, experiments, and wargames, it should work closely with AFMC/A3X in developing this rough draft into a comprehensive instruction.

Air Force Instruction 10-401, *Air Force Operations Planning and Execution*, December 7, 2006

The purpose of AFI 10-401 is to provide an overview of the joint planning process and the interrelationships of the associated national-level systems that produce national security policy, military strategy, force and sustainment requirements, and plans. The four major interrelated systems affecting the development of joint operational plans are (1) the National Security Council System (NSCS); (2) the Joint Strategic Planning System (JSPS); (3) the Planning, Programming, Budgeting, and Execution (PPBE) Process; and (4) the Joint Operation Planning and Execution System (JOPES). This instruction provides very detailed planning guidance.

The logistics supplement to the War and Mobilization Plan Volume 1 (WMP-1) provides guidance for directing LSA as directed by Chairman of the Joint Chiefs of Staff Memorandum (CJCSM) 3122.03B and Chairman of the Joint Chiefs of Staff Instruction (CJCSI) 3110.03C, *Logistics Supplement to the Joint Strategic Capabilities Plan (JSCP)*. The LSA anticipates combat support challenges and resolves them before they become showstoppers. The LSA addresses the areas of materiel, infrastructure, logistics support forces, and lift in detail. It identifies deficiencies, assesses the risk or impact on operations and any known get-well dates or alternative solutions, and assigns a level of risk associated with the deficiency.

Only significant deficiencies requiring external assistance need addressing. Other Air Force providers of combat forces, resources, and capabilities also provide their assessment of sustainability to the A4. The entire intent of the LSA is to provide a broad assessment of the key combat logistics support and enabler capabilities required to execute the CCDR's planned operation.

It is the responsibility of functional capability managers and planners at all levels to analyze and review WMP-1 guidance for their respective functional capabilities. Functional capability managers will work closely with Air Staff to ensure compliance with guidance, resolve any contentious issues, and ensure the most effective management of forces.

COMAFFOR Senior Staff Course (CSSC). CSSC is a mentored seminar for Air Force colonels exercising executive responsibilities, recommending force application and movement, maintaining situational awareness, and developing responsive COAs and adaptive plans in C2 organizations above base level. COMAFFOR Special and A-Staff Directors are specifically targeted, as are Air Force colonels supporting COCOM staffs. In a seminar setting, senior officers examine critical COMAFFOR and COCOM situations and lessons learned through case studies and mentor interaction. AF/A4RC is the Air Staff CSSC sponsor.

For sustainment planning, AFMC/A4R validates all logistics planning factors developed by Air Force and other DoD organizations. Headquarters Air Force, Deputy Chief of Staff for Logistics, Installations, and Mission Support (AF/A4/7) reviews these planning factors to ensure they are consistent with policy guidance.

AFMC responsibilities are detailed with no mention of the Air Force Sustainment Center. The Chief, Logistics Readiness Division, Director of Logistics and Sustainment is the Air Force central manager for LSA development, validation and dissemination of wartime resupply planning factors. This office provides planners with approved wartime resupply planning factors for determining logistics support strategic lift requirements based on force structure, the length of generation, and other scenario conditions.

Specifically, AFMC/A4R does the following:

- provides functional guidance relative to the use, development, computation, validation, and management of wartime resupply planning factors
- coordinates wartime resupply planning factor policy decisions
- keeps affected agencies informed of proposed planning factor program changes
- maintains liaison with the respective Air Force collateral managers of classes and subclasses of supply and other military services, as well as DoD agencies involved in the development and use of wartime resupply planning factors
- documents lessons learned and maintains audit trails on methods, rationale, and data sources used for the development of planning factors
- functions as the lead Air Force activity for updating wartime resupply planning factors
- validates all Air Force wartime resupply planning factors prior to their inclusion in the Logistics Factors File (LFF) in JOPES

- transmits sustainment planning data for the Air Force Class IX supply (less medical peculiar repair parts)
- develops new methods and automatic data processing system capabilities to improve data collection and computation of wartime resupply planning factors
- interacts with other military services, DoD organizations, Air Force MAJCOMs, and agencies for data exchange to support existing and improved methods for sustainment planning factor development
- acts as the focal point for developing the capability to link sustainment requirements with wholesale item asset availability
- verifies consumption factor updates to the JOPES LFF.

ACS sustainment planning is a crucial element of crisis action and contingency planning. The Air Force accomplishes this planning by means of an LSA. LSA is an analytical process used to predict ACS operational capability requirements, gaps, and priorities. The process and methodology support Defense Planning Guidance (DPG) and major theater OPLAN assessments, crisis action planning, and supplemental budgeting estimates. AFMC/A4R validates all logistics planning factors developed by Air Force and other DoD organizations. AF/A4/7 reviews these planning factors to ensure they are consistent with policy guidance, ACS concept of operations (CONOPS) objectives, and Capabilities Review and Risk Assessment scenarios and priorities. This assessment provides a broad assessment of key ACS support and enabler capabilities required to execute the DPG and the COCOM's plans. *As a general rule, the Air Force uses the supported component headquarters' directorate of logistics, or equivalent, as its agent for analysis.*

The LSA is accomplished in accordance with JSCP, CJCSI 3110.03C, and CJCSM 3122.03B. The LSA must be submitted to the supported commander for inclusion in the theater LSA for the OPLAN. Air Force supporting commands are also required to accomplish an LSA and submit results to the supported COMAFFOR. The LSA addresses the four pillars of ACS sustainability (materiel, infrastructure, expeditionary combat support (ECS) forces, and lift). It highlights deficiencies and their associated risk to supporting the warfighting air component.

Air Force Instruction 10-401, *Air Force Operations Planning and Execution,* December 7, 2006, Air Force Materiel Command Supplement, June 7, 2006, Incorporating Through Change 2, *Air Force Operations Planning and Execution*, July 29, 2009

This instruction embeds AFMC–unique situations and aligns the Command with AFI 10-401. It has a significant amount of inclusions and incorporates AFMC Guidance Memorandum 10-01-2008 for managing and conducting AEF processes and tasks. This annotation will only add applicable AFMC-unique inputs that are not written in the AFI version of this instruction.

In support of crisis action planning, this instruction identifies AFMC Command Center (OPSO/A3XC) as the OPR for receipt and validation of higher headquarters' Planning and Execution Orders and AFMC/A3X (AFMC CAT/A3-Deployment Cell) as the Command OPR for receipt and action on the Joint Chiefs of Staff (JCS)–directed execution of OPLANs, CONPLANs, DEPORDs, and operations orders. AFMC/A3X will prepare and maintain procedures for operation of the AFMC/A3-Deployment Cell.

In support of crisis action planning, all AFMC A-staff, functional directorates, installations, DRUs, and geographically separated units (GSUs) will have primary and alternate POCs trained and proficient in contingency and crisis action planning for their function. Contingency and crisis action procedures must be periodically exercised during joint and unilateral command post exercises and field training exercises to ensure the required capability is available. AFMC participation in any exercise involving crisis action planning should be consistent with real-world processes.

AFMC/A3X is the Command OPR for unified and specified command plans and reviews all plans for impact on AFMC. Further, the Installation Commander is responsible for ensuring their plans are reviewed on a regular basis. All plans and implementing procedures will be reviewed every 12 months. Each AFMC Commander must delineate, in detail, the actions to be performed by each organization involved in supporting emergency tasks for which he or she is responsible.

AFMC/A3X is the focal point for coordinating all plans (whether produced by Headquarters AFMC or another AFMC entity) with other MAJCOMs. The AFMC Exercise Program (AFMC/A3X), in conjunction with the AFMC IG and other OPR functions in the MAJCOM, including exercise POCs for AFMC installations, centers, and GSUs/DRUs, will coordinate an annual process whereby all AFMC-scheduled exercises and AFMC-scheduled exercise-related activities (for example, readiness exercises, IG inspections involving events termed exercises, experiments, wargames, demonstrations [capabilities/technology/other], and termed exercises) are—to the maximum extent practical—synchronized within the Command and with the AEF battle rhythm. The overall process is detailed in AFMCI 10-204, *AFMC Exercise-Related Activities and Support*.

Air Force Instruction 10-404, *Base Support and Expeditionary Site Planning*, March 9, 2004

This instruction implements AFPD 10-4, *Operations Planning*, and provides for the preparation of base support plans (BSPs), expeditionary site plans (ESPs), and the accomplishment of contingency site surveys across the spectrum of Air Force operations for deliberate and crisis action planning and execution. It describes what is needed to translate and integrate operational requirements into ACS and ECS at employment sites to create and sustain operations. This revision integrates the In-Garrison Expeditionary Site Plan (IGESP) and the Expeditionary Site Survey Process (ESSP) into the plan.

AFI 10-404 states the objectives of IGESP and ESP as determining capabilities and applying them to contingency operations. The ESSP is a subset of the overall expeditionary site planning process, which is the foundation for Air Force expeditionary operations. It provides the detailed information required by planners at all levels—strategic, operational, and tactical. Whether they are developing the air campaign, the aircraft basing plan supporting the air campaign, or preparing to deploy a unit forward to execute the plan, all planners require similar information to begin planning. Part I of the IGESP and ESP identifies the resources and capabilities of a location by functional capability and is the focus of the expeditionary site survey. For contingency requirements, Part II of the plan allocates the resources identified in Part I, assesses the ability to support the operation, and identifies limiting factors (LIMFACs). IGESPs are primarily developed for locations with a permanent Air Force presence and are fully developed by the collaborative planning efforts of many functional experts with a deliberate planning time line. ESPs are chiefly associated with locations without a permanent Air Force presence and may contain only the minimum data necessary to make initial beddown decisions (quick reaction site survey information in Part I). ESPs may be developed within short time frames to meet contingency needs without full staffing or coordination.

Planners use the Logistician's Contingency Assessment Tools (LOGCAT), a suite of standard systems tools that enables automated, employment-driven, ACS planning. LOGCAT supports the expeditionary site planning process by accurately and rapidly identifying resources and combat support requirements at potential employment locations, providing beddown capability analysis and LIMFAC identification, and facilitating force tailoring decisions to reduce the overall deployment footprint. LOGCAT consists of three components that are mandated for use when they are available at all levels of command. The baseline planning data for IGESP/ESP development is (1) COCOMs and supporting OPLANS and CONPLANS, (2) TPFDDs including all-service data, (3) wartime aircraft activity reports (WAARs), (4) WRM authorization documents, and (5) contingency-in-place requirements. Planners and surveyors must be able to take advantage of the DoD Communications Network, the Global Command and Control System (GCCS), and Global Combat Support System (GCSS) infrastructures.

AFI 10-404 could include two products to enable the planning process: (1) the Strategic Tool for the Analysis of Required Transportation (START) model and (2)

FAM. START determines early-stage manpower and equipment deployment requirements. It is a preliminary requirements TPFDD generator that COCOMs could use to generate a list of required UTCs to support a user-specified operation. The UTC requirements are a function of rules (for example, the number and type of aircraft beddown, beddown conditions, and threat conditions). A fully developed version of this tool could enable the kind of quick planning processes early in the contingency planning cycle that could prove to be useful to the Air Force in both deliberate and crisis action planning. FAM computes aggregate demands and supplies a list of required UTCs. Demands for UTCs can come from TPFDDs, if available, or from other estimates of the forces required to execute a particular operation, and the UTCs come from AEF Reporting Tool (ART). Given an input TPFDD (or other demand list) and an input ART file, FAM first identifies if the demand file contains any faulty UTCs. In the event that faulty UTCs are identified (that is, not green), subject-matter experts can review ART and the commander's comment for UTCs to determine if a UTC could become fully operational according to the ART reporting standards or if the demand needs to be changed or deleted. FAM does not address sourcing decisions.

AFMC Guidance and Policy for Material Surge and Plan 70, January 27, 2009 (Draft)

This policy will be used in conjunction with Air Force and AFMC policy and is developed in support of AFMC Plan 70. This plan provides a mechanism to request and obtain additional depot support/resources to meet increased peacetime and/or contingency requirements. Plan 70 outlines and defines the processes used to plan and manage the transition from peacetime materiel support levels to those required to maintain both contingency and wartime support levels. Materiel support may consist of any combination of commodities, engines, and/or aircraft. It satisfies AFI 21-102 requirements for AFMC to develop and maintain a surge CONPLAN, contains guidance and procedures for the Air Force Sustainment Center and ALCs to develop surge plans, and defines the process that the Sustainment Center and ALCs will use to plan and manage the transition from peacetime to contingency support levels. Notably, the ALCs should include procedures for surging exchangeable parts, both within and outside the Execution and Prioritization of Repair Support System (EXPRESS); accelerating/compressing aircraft in depot maintenance, as required; and coordinating surge plans with other ALCs.

However, there is an absence of command direction that outlining how the ALCs, centers, and directorates will plan for OPLAN and contingency planning and execution, exercises, experiments, and wargames.

Warner Robins Air Logistics Center (WR-ALC) and 638th Supply Chain Management Group, *Surge Contingency Plan 70*, December 2009

This plan provides policy and guidelines for the surge production of exchangeables, the acceleration/compression of aircraft during contingency situations, and acquisition surge/acceleration operations in support of contingency operations and AEF steady-state requirements. During any contingency, and for the duration of steady-state requirements, the ALC will ensure the highest depot-level production possible to meet the needs of operational forces and national objectives forwarded from the AFMC Crisis Action Team (CAT), which helps estimate contingency support activity and the possible implementation of surge. The CAT forwards orders from the JCS (which serve as key milestones in the contingency execution process and provide updated information on timing, taskings, etc.). The Strategic Planning Branch (WR-ALC/XPTS) maintains the WR-ALC Staff Control Center (SCC), which functions as a unit control center and reports to the Robins Installation Control Center (ICC), and maintains the WR-ALC Materiel Control Center (MCC). During surge, the MCC will function as the WR-ALC commander's C2 hub for Wartime Materiel Support (WMS) issues; support surge operations, as presented in Annex C; and other portions of this plan, as tasked. Surge can be directed by the AFMC/CC, WR-ALC/CC, and/or the 638th Supply Chain Management Group (638 SCMG) Director in response to increased requirements due to a steady-state or wartime contingency.

Surge of exchangeable requirements in support of steady-state and contingency operations applies to stock-fund managed exchangeable items; some processes may also apply to items managed by other systems. This policy provides repair-cycle procedures for items controlled by the Depot Repair Enhancement Program (DREP) and the Contract Repair Process (CRP) and driven by EXPRESS. It also provides specific management procedures for the 638 SCMG, 330th Aircraft Sustainment Wing, 402nd Maintenance Wing, Financial Management Directorate (WR-ALC/FM), the materiel management (MM) and item management (IM) functions, and the production management functions within the Source of Repair (SOR) Groups. Included are instructions for the WR-ALC and 638 SCMG Exchangeable Surge POC, the Group/Squadron Exchangeable Surge POC, the Exchangeable Surge Committee, and the DREP and CRP teams outlining specific responsibilities for functional capabilities and team personnel involved in the repair process. The DREP and CRP team processes include all WR-ALC and 638 SCMG Source of Supply (SOS) Groups, as well as the SOR, contracting, supply, preservation/packaging, and material movement functions for both EXPRESS and non-EXPRESS items. Exchangeable surge procedures are generally initiated by information contained in JCS orders indicating or directly requesting a need for increased production or expedited delivery.

Aircraft production surge is in response to a formal customer request. Aircraft acceleration/compression can be directed by the AFMC/CC or WR-ALC/CC. Surge is used by AFMC to accelerate production during depot maintenance when the owning commands require increased aircraft support. A surge in aircraft depot maintenance is accomplished through the acceleration or compression modes of production. AFMC depots will surge aircraft in response to customer requests after a cost analysis has been performed and funds have been made available. It is the responsibility of the owning command to identify any increased aircraft demands necessary to successfully complete contingency activity.

Accelerated acquisition of a new program is in response to specific wartime requirements. Acquisition surge is the acceleration of an ongoing program to meet wartime requirements.

Oklahoma City ALC (OC-ALC), *Surge Contingency Plan 70*, May 2009

This plan provides policy and guidance for planning and executing depot level maintenance surge activities in support of contingency operations and Air, Space, and Cyber Space Expeditionary Force steady-state requirements. It provides a mechanism for requesting and obtaining additional depot support/resources to meet increased peacetime and/or contingency requirements. The plan outlines and defines the processes used to plan and manage the transition from peacetime materiel support levels to those required to maintain both contingency and wartime support levels. Materiel support may consist of any combination of commodities, engines, and/or aircraft. The plan provides guidance for the planning and implementation of acquisition acceleration/surge to provide materiel support to contingency operations. *Accelerated acquisition* is a new program in response to specific wartime requirements. *Acquisition surge* is the acceleration of an ongoing program to meet wartime requirements. The plan also provides exchangeable surge policy and procedures to meet both EXPRESS and non–EXPRESS-driven exchangeable requirements in support of steady-state and contingency operations. Although this policy applies to stock-fund managed exchangeable items, processes for other items are also addressed.

This plan is developed in support of AFMC Plan 70. OC-ALC and 448 SCMW must provide logistics support to ensure customers have the capability to integrate and adapt operations that achieve strategic and tactical effects in a total joint force environment. AFMC forwards contingency information and JCS orders (from the AFMC/CAT), which will assist in estimating contingency support requirements and implementing surge. The CAT coordinates contingency support operations throughout the command and acts as the single headquarters' focal point for incoming and outgoing contingency communications with higher headquarters, lateral contingency response staffs, and the AFMC center-level contingency staffs. The CAT acquires and disseminates key information that assists single managers in supporting the contingency. The 72 Airbase Wing ICC will receive/analyze/distribute information.

Ogden ALC (OO-ALC), *Depot Level Wartime Material Support Contingency Plan 70*, June 29, 2009

This plan directs acceleration and surge operations at OO-ALC in support of the Air Force, DoD, AFMC, and JCS OPLANs, including the acceleration or compression of aircraft in OO-ALC facilities for programmed depot maintenance or modification and acquisition acceleration in response to a contingency, emergency, or exercise. This document is effective for planning and implementation directed by the OO-ALC/CC or a higher authority. It provides guidance for the OO-ALC implementation of AFI 63-114, *Rapid Response Process* and *Headquarters AFMC Plan 70*, Surge Contingency Plan. It provides the OO-ALC commander with the flexibility to accelerate the fielding of critical systems and implement aircraft surge procedures to meet theater-specific wartime needs, including support of forces in conflict or crisis situations. In the event of acquisition acceleration or surge/compression operations, OO-ALC Plan 70 delineates responsibilities, lines of communication, and the actions to be taken to ensure continued readiness to rapidly provide the highest level of depot production and materiel support possible during any contingency situation, commensurate with operational force needs and national objectives. Upon request from a supported MAJCOM and/or direction from AFMC/A4 for surge, OO-ALC/CC executes Plan 70 as required. Air Force WMP-1 war planning assumptions apply.

Operational Lead Commands (Force Providers) are responsible for identifying the aircraft needed to meet contingency taskings. Commands are required to submit surge requests for aircraft to the respective system program director (SPD) and Headquarters AFMC. Prior to a request, early indications may come from JCS Warning or Planning Orders. Based on these indications, the SPD should begin an assessment of the ability to support a surge. OO-ALC will respond to aircraft surge requests using acceleration or compression sustainment measures. The acquisition acceleration process is initiated when the SPD is notified of a critical need that cannot be met by existing fielded systems. Initial notification of a critical need may be provided formally or informally to the SPD by a supported MAJCOM or an acquisition authority.

The OO-ALC WMS tasking process may require the activation of a C2 element, which is a process of Hill Air Force Base Plan 8. The ICC, or portions of it, may be activated to respond to a WMS request depending on the scope of the support directed by the OO-ALC/CC or higher headquarters. OO-ALC may be the lead ALC to support a surge request or it may support a lateral surge requirement. WMS response options will take priority over OO-ALC peacetime efforts.

Exchangeable surge policy and procedures apply to Defense Working Capital Fund–managed exchangeable items, including items repaired under the DREP and CRP. Tasking orders will usually include some type of military activity/buildup at specified locations. Only the OO-ALC/CC, with authorization from Headquarters AFMC or JCS orders dictated by events, or upon request of the supply chain manager or SPD, has the authority to direct a surge. With this authorization, the depots may expend additional work forces (extended shifts, weekends, second and/or third shifts), reassign personnel to shops with the highest-priority workloads, increase shop capacity, procure additional materiel, and/or spend additional funds for contractor support to meet the increased contingency demands. The intent is to provide CCs with a variety of options for responding to a contingency.

Air Force Global Logistics Support Center, *Surge Contingency Plan 70*, May 26, 2010 (Draft)

This plan supports AFMC's Plan 70. It provides a mechanism to request and obtain additional depot support/resources to meet increased peacetime and/or contingency requirements. It outlines and defines the processes used to plan and manage the transition from peacetime materiel support levels to those required to maintain both contingency and wartime support levels. Materiel support may consist of any combination of commodities, engines, and/or aircraft. This is a very important document that addresses contingency operations but does not address OPLAN support planning. Suggested inputs include:

- Define how the 591st Supply Chain Management Group will perform proactive assessments to identify potential problem items at the beginning of contingency planning and add a requirement for Weapon Systems Management Information System and PC-Aircraft Sustainability Model (PC-ASM) assessments. Describe this analysis and participation with C-NAF staff in the formulation of CONPLANs or the evaluation of OPLANs.
- State that ALC OPLAN 70s should also address official OPLAN assessments and have a draft plan ready to meet OPLANs, if executed.
- Define and specify relationships with C-NAFs, the OSC ACS reachback cell, AF/A4/7, and AF/A3/5 to get priorities to input into EXPRESS.

- Specify how the Air Force Sustainment Center assists all the ALCs and state the authority of EXPRESS data to authorize the Sustainment Center to direct repairs at all the ALCs.
- Develop procedures for the critical item programs to follow in support of OPLAN LSAs.

AFMC/A8XW, *Wargaming Integration Office Charter*, March 2011 (Draft)

This charter states that the mission of the AFMC Wargaming Integration Office (AFMC/A8XI) is to provide command-level oversight of wargames and provide a process for utilizing wargame feedback to assist AFMC planning and programming. With the most recent changes in the planning and programming cycles, wargame results can provide valuable guidance to both the Air Force Strategic Planning System and Air Force Corporate Structure processes. Specifically for AFMC, the ACS Core Function Lead Integrator (CFLI) can utilize feedback in the planning process to influence the ACS Core Function Master Plan and POM. Wargame results can also influence and provide direction for AFMC's role in the remaining 11 service core functions and provide for additional impact (funding, doctrine, etc.) to assist or emphasize a particular capability or concept.

This charter defines the proposed AFMC/A8XI roles and responsibilities and identifies the major collaborators required for successful AFMC participation in Title X wargames. In addition, this charter describes the overall AFMC wargaming process and the planned way ahead for AFMC/A8XI. Two of the major focus areas of the office will be to ensure that wargaming results and feedback are integrated into the overall AFMC ACS planning and programming cycles and that the results and feedback are channeled back to the respective agencies for their internal influence, prioritization, and action.

Bibliography

"A New Vision for Global Support, C2 Combat Support," *Air Force Journal of Logistics*, Vol. XXVII, No. 2, Summer 2003.

Boyd, John R., "A Discourse on Winning and Losing," Maxwell AFB, Alabama: Air University Library, Document No. M-U4397, unpublished collection of briefing slides, August 1987.

Brown, Kendall, "The Role of Air Force Civil Engineers in Counterinsurgency Operations," *Air & Space Power Journal*, Summer 2008.

Defense Acquisition University, *Systems Engineering Fundamentals*, Supplement 5-B: IDEF0, Fort Belvoir, Va.: Defense Acquisition University Press, January 2001.

Feinberg, Amatzia, Eric Peltz, James A. Leftwich, Robert S. Tripp, Mahyar A. Amouzegar, Russell Grunch, John G. Drew, Tom LaTourrette, and Charles Robert Roll, Jr., *Supporting Expeditionary Aerospace Forces: Lessons from the Air War Over Serbia*, Santa Monica, Calif.: RAND Corporation, MR-1263-AF, 2002, not available to the general public.

Feinberg, Amatzia, Hyman L. Shulman, Louis W. Miller, and Robert S. Tripp, *Supporting Expeditionary Aerospace Forces: Expanded Analysis of LANTIRN Options*, Santa Monica, Calif.: RAND Corporation, MR-1225-AF, 2001. As of August 19, 2013:
http://www.rand.org/pubs/monograph_reports/MR1225.html

Fox, Christine, Cost Assessment and Program Evaluation (CAPE), "DoD Efficiency Decisions," briefing, August 9, 2010.

Gabreski, Terry L., et al., "Command and Control Doctrine for Combat Support: Strategic- and Operational-Level Concepts for Supporting the Air and Space Expeditionary Force," *Air and Space Power Journal*, Spring 2003.

Hanrahan, Robert P., *The IDEF Process Modeling Methodology*, Software Technology Support Center, June 1995.

Joint Publication 3-0, *Joint Operations*, Washington, D.C.: Joint Chiefs of Staff, September 17, 2006, Incorporating Change 1, February 13, 2008.

Joint Publication 5-0, *Joint Operation Planning*, Washington, D.C.: Joint Chiefs of Staff, December 26, 2006.

Kent, Glenn, *A Framework for Defense Planning*, Santa Monica, Calif.: RAND Corporation, R-3721-AF/OSD, 1989. As of August 19, 2013: http://www.rand.org/pubs/reports/R3721.html

Leftwich, James, Robert S. Tripp, Amanda B. Geller, Patrick Mills, Tom LaTourrette, Charles Robert Roll, Jr., Cauley Von Hoffman, and David Johansen, *Supporting Expeditionary Aerospace Forces: An Operational Architecture for Combat Support Execution Planning and Control*, Santa Monica, Calif.: RAND Corporation, MR-1536-AF, 2002. As of August 19, 2013: http://www.rand.org/pubs/monograph_reports/MR1536.html

Lewis, Leslie, James A. Coggin, and Charles Robert Roll, Jr., *The United States Special Operations Command Resource Management Process: An Application of the Strategy-to-Tasks Framework*, Santa Monica, Calif.: RAND Corporation, MR-445-A/SOCOM, 1994. As of August 19, 2013: http://www.rand.org/pubs/monograph_reports/MR445.html

Lynch, Kristin F., John G. Drew, Robert S. Tripp, and Charles Robert Roll, Jr., *Supporting Air and Space Expeditionary Forces: Lessons from Operation Iraqi Freedom*, Santa Monica, Calif.: RAND Corporation, MG-193-AF, 2005. As of August 19, 2013: http://www.rand.org/pubs/monographs/MG193.html

Lynch, Kristin F., John G. Drew, Robert S. Tripp, Daniel M. Romano, Jin Woo Yi, and Amy L. Maletic, *An Operational Architecture for Improving Air Force Command and Control Through Enhanced Agile Combat Support Planning, Execution, Monitoring, and Control Processes*, Santa Monica, Calif.: RAND Corporation, RR-261-AF, 2014. As of June 2014: http://www.rand.org/pubs/research_reports/RR261.html

Lynch, Kristin F., and William A. Williams, *Combat Support Execution Planning and Control: An Assessment of Initial Implementations in Air Force Exercises*, Santa Monica, Calif.: RAND Corporation, TR-356-AF, 2009. As of August 19, 2013: http://www.rand.org/pubs/technical_reports/TR356.html

McFarren, Michael, informational briefing, SAF/US(M), May 2010.

McGarvey, Ronald G., James M. Masters, Louis Luangkesorn, Stephen Sheehy, John G. Drew, Robert Kerchner, Ben D. Van Roo, and Charles Robert Roll, Jr., *Supporting Air and Space Expeditionary Forces: Analysis of CONUS Centralized Intermediate Repair Facilities*, Santa Monica, Calif.: RAND Corporation, MG-418-AF, 2008. As

of August 19, 2013:
http://www.rand.org/pubs/monographs/MG418.html

McGarvey, Ronald G., Robert S. Tripp, Rachel Rue, Thomas Lang, Jerry M. Sollinger, Whitney A. Conner, and Louis Luangkesorn, *Global Combat Support Basing: Robust Prepositioning Strategies for Air Force War Reserve Materiel*, Santa Monica, Calif.: RAND Corporation, MG 902-AF, 2010. As of August 19, 2013: http://www.rand.org/pubs/monographs/MG902.html

Mills, Patrick, John G. Drew, John A. Ausink, Daniel M. Romano, and Rachel Costello, *Balancing Agile Combat Support Manpower to Better Meet the Future Security Environment*, Santa Monica, Calif.: RAND Corporation, RR-337-AF, 2014. As of June 2014:
http://www.rand.org/pubs/research_reports/RR337.html

Mills, Patrick, Ken Evers, Donna Kinlin, and Robert S. Tripp, *Supporting Air and Space Expeditionary Forces: Expanded Operational Architecture for Combat Support Execution Planning and Control*, Santa Monica, Calif.: RAND Corporation, MG-316-AF, 2006. As of August 19, 2013:
http://www.rand.org/pubs/monographs/MG316.html

Peltz, Eric, Hyman L. Shulman, Robert S. Tripp, Timothy Ramey, and John G. Drew, *Supporting Expeditionary Aerospace Forces: An Analysis of F-15 Avionics Options*, Santa Monica, Calif.: RAND Corporation, MR-1174-AF, 2000. As of August 19, 2013:
http://www.rand.org/pubs/monograph_reports/MR1174.html

Schrader, John Y., Leslie Lewis, William Schwabe, Charles Robert Roll, Jr., and Ralph Suarez, *USFK Strategy-to-Task Resource Management: A Framework for Resource Decisionmaking*, Santa Monica, Calif.: RAND Corporation, MR-654-USFK, 1996. As of August 19, 2013:
http://www.rand.org/pubs/monograph_reports/MR654.html

Snyder, Don, and Patrick Mills, *Supporting Air and Space Expeditionary Forces: A Methodology for Determining Air Force Deployment Requirements,* Santa Monica, Calif.: RAND Corporation, MG-176-AF, 2004. As of January 7, 2014:
http://www.rand.org/pubs/monographs/MG176.html

Thaler, David E., *Strategies to Tasks: A Framework for Linking Means and Ends*, Santa Monica, Calif.: RAND Corporation, MR-300-AF, 1993. As of August 19, 2013:
http://www.rand.org/pubs/monograph_reports/MR300.html

Tripp, Robert S., Lionel A. Galway, Paul Killingsworth, Eric Peltz, Timothy Ramey, and John G. Drew, *Supporting Expeditionary Aerospace Forces: An Integrated Strategic Agile Combat Support Planning Framework*, Santa Monica, Calif.: RAND Corporation, MR-1056-AF, 1999. As of August 19, 2013: http://www.rand.org/pubs/monograph_reports/MR1056.html

Tripp, Robert S., Lionel A. Galway, Timothy L. Ramey, Mahyar A. Amouzegar, and Eric Peltz, *Supporting Expeditionary Aerospace Forces: A Concept for Evolving to the Agile Combat Support/Mobility System of the Future*, Santa Monica, Calif.: RAND Corporation, MR-1179-AF, 2000. As of August 19, 2013: http://www.rand.org/pubs/monograph_reports/MR1179.html

Tripp, Robert S., Kristin F. Lynch, John G. Drew, and Edward W. Chan, *Supporting Air and Space Expeditionary Forces: Lessons from Operation Enduring Freedom*, Santa Monica, Calif.: RAND Corporation, MR-1819-AF, 2004. As of August 19, 2013: http://www.rand.org/pubs/monograph_reports/MR1819.html

Tripp, Robert S., Kristin F. Lynch, John G. Drew, and Robert G. DeFeo, *Improving Air Force Command and Control Through Enhanced Agile Combat Support Planning, Execution, Monitoring, and Control Processes*, Santa Monica, Calif.: RAND Corporation, MG-1070-AF, 2012. As of August 19, 2013: http://www.rand.org/pubs/monographs/MG1070.html

Tripp, Robert S., Kristin F. Lynch, Charles Robert Roll, Jr., John G. Drew, and Patrick Mills, *A Framework for Enhancing Airlift Planning and Execution Capabilities within the Joint Expeditionary Movement System,* Santa Monica, Calif.: RAND Corporation, MG-377-AF, 2006. As of August 19, 2013: http://www.rand.org/pubs/monographs/MG377.html

U.S. Air Force, "Command and Control Service Core Function, 2010-2030," undated draft, not available to the general public.

U.S. Air Force, Air Force Manual 13-220, *Deployment of Airfield Operations*, May 13, 1997.

U.S. Air Force, Air Force Doctrine Document 1, *Air Force Basic Doctrine*, November 17, 2003.

U.S. Air Force, Air Force Instruction 10-404, *Base Support and Expeditionary Site Planning*, March 9, 2004.

U.S. Air Force, Air Force Instruction 13-204, *Functional Management of Airfield Operations*, January 10, 2005a.

U.S. Air Force, Air Force Instruction 13-203, *Air Traffic Control*, November 30, 2005b.

U.S. Air Force, *Air Force Forces Command and Control Enabling Concept*, Change 2, May 25, 2006a.

U.S. Air Force, Program Action Directive 06-09, *Implementation of the Chief of Staff of the Air Force Direction to Establish an Air Force Component Organization*, November 7, 2006b.

U.S. Air Force, Air Force Instruction 10-401, *Air Force Operations Planning and Execution*, December 7, 2006c.

U.S. Air Force, *Agile Combat Support CONOPS*, November 15, 2007.

U.S. Air Force, Air Force Instruction 13-213, *Airfield Management*, January 29, 2008a.

U.S. Air Force, *Agile Combat Support Command and Control (ACS C2) Supporting CONOPS*, November 15, 2008b.

U.S. Air Force, *Oklahoma City Air Logistics Center Surge Contingency Plan,* OC-ALC Plan 70, May 2009a.

U.S. Air Force, *Ogden Air Logistics Center Depot Level Wartime Material Support Contingency Plan 70,* OO-ALC Plan 70, June 29, 2009b.

U.S. Air Force, Air Force Instruction (AFI) 10-401, *Air Force Operations Planning and Execution*, December 7, 2006, Air Force Materiel Command (AFMC) Supplement, June 7, 2006 Incorporating Through Change 2, July 29, 2009c, *Air Force Operations Planning and Execution.*

U.S. Air Force, *Warner Robins Air Logistics Center & 638th Supply Chain Management Group Surge Contingency Plan 70*, December 2009d.

U.S. Air Force, *Agile Combat Support Core Function Master Plan 2010*, draft version as of January 19, 2010a.

U.S. Air Force, Program Action Directive 10-02, *Implementation of the Chief of Staff of the Air Force Direction to Restructure Command and Control of Component Numbered Air Force*, June 2, 2010b.

U.S. Air Force, Air Combat Command (ACC), Air Force Command and Control Integration Center (AFC2IC), Joint Expeditionary Force Experiment (JEFX) 2011, JEFX 11-1, *Agile Logistics Evaluation Experiment (ALEX) Final Report,* April 25, 2011.

U.S. Air Force Global Logistics Support Center, Surge Contingency Plan 70 (Draft), May 26, 2010.

U.S. Air Force Materiel Command, *Guidance & Policy for Materiel Surge & Plan 70*, January 27, 2009.

U.S. Air Force Materiel Command, Air Force Materiel Command Supplement 1 to Air Force Instruction 10-401 (AFI 10-401 AFMCSUP I), *Air Force Operations Planning and Execution*, July 29, 2010a.

U.S. Air Force Materiel Command, Air Force Materiel Command Instruction 10-204, *AFMC Exercise Program*, August 31, 2010b.

U.S. Army, Field Manual 5-0, *The Operations Process*, March 2010.

U.S. Department of Defense, "DoD Architecture Framework Overview," PowerPoint briefing, Alessio Mosto, May 2004.

U.S. Department of Defense, *DoD Architecture Framework*, Version 1.5, Volume I: *Definitions and Guidelines*, April 23, 2007a.

U.S. Department of Defense, *Guidance for Employment of the Force (GEF)*, draft, August 31, 2007b, not available to the general public.

U.S. Department of Defense, *Guidance for Development of the Force,* April 2008a, not available to the general public.

U.S. Department of Defense, *Global Force Management Implementation Guidance (GFMIG) FY 08–09*, June 4, 2008b, not available to the general public.

U.S. Department of Defense, *DoD Architecture Framework*, Version 2.0, Volume I: *Introduction, Overview, and Concepts (Manager's Guide)*, May 28, 2009.